About Island Press

Island Press is the only nonprofit organization in the United States whose principal purpose is the publication of books on environmental issues and natural resource management. We provide solutions-oriented information to professionals, public officials, business and community leaders, and concerned citizens who are shaping responses to environmental problems.

In 1994, Island Press celebrated its tenth anniversary as the leading provider of timely and practical books that take a multidisciplinary approach to critical environmental concerns. Our growing list of titles reflects our commitment to bringing the best of an expanding body of literature to the environmental community throughout North America and the world.

Support for Island Press is provided by Apple Computer, Inc., The Bullitt Foundation, The Geraldine R. Dodge Foundation, The Energy Foundation, The Ford Foundation, The W. Alton Jones Foundation, The Lyndhurst Foundation, The John D. and Catherine T. MacArthur Foundation, The Andrew W. Mellon Foundation, The Joyce Mertz-Gilmore Foundation, The National Fish and Wildlife Foundation, The Pew Charitable Trusts, The Pew Global Stewardship Initiative, The Rockefeller Philanthropic Collaborative, Inc., and individual donors.

THE
FLORIDA
PANTHER

THE
FLORIDA
PANTHER

Life and Death of a Vanishing Carnivore

DAVID S. MAEHR

ISLAND PRESS
Washington, D.C.   Covelo, California

ISLAND PRESS is a trademark of The Center for Resource Economics.
All figures are by David S. Maehr, except figure 10.2, which is reprinted by permission of E. Darrell Land.

Library of Congress-in-Publication Data

Maehr, David S., 1955–
    The Florida panther : life and death of a vanishing carnivore /
David S. Maehr.
        p.   cm.
    Includes bibliographical references.
    ISBN 1-55963-506-1 (cloth). — ISBN 1-55963-507-x (pbk.)
        1. Florida panther.   2. Wildlife conservation—Florida.
    I. Title.
    QL737.C23M24   1998
    599.75'24—dc21   97-14844
                                                      CIP

Printed on recycled, acid-free paper ∞ ✪

Manufactured in the United States of America

10 9 8 7 6 5 4 3 2 1

CONTENTS

ACKNOWLEDGMENTS

Creating this book was often a lonely endeavor—poring over field notes, data sheets, computer files, pencil sketches, reports, and waning memories. In doing so, however, I discovered I was no longer the same person who had stumbled into the swamps and politics of the Florida panther. When I reemerged, I believe I was a better person for the experience and that my colleagues in the field were instrumental in this transformation. They were also instrumental in defining the animal that we came to know so well. Darrell Land, Roy McBride, Walt McCown, Melody Roelke, and Jayde Roof each added to the richness of our combined experience with their own brands of humor, professionalism, dedication, and individual quirks. Without them, the panther would still be a great mystery.

This rich milieu of research and personal interaction was made possible by seldom-seen administrators in Gainesville and Tallahassee. Brad Gruver, Tommy Hines, and Tom Logan juggled the Florida Game and Fresh Water Fish Commission funding and policy headaches that were constant elements of the work. Todd Logan, former manager of the Florida Panther National Wildlife Refuge and latecomer to the days of expanding research efforts, as well as Sonny Bass of Everglades National Park, demonstrated that state and federal agencies can work closely and effectively together in reaching frequently contentious goals. That some of them are not as closely involved today is the panther's misfortune.

This book would not have evolved to completion without the interest and dedication of my editor, Barbara Dean. Her diplomacy made the necessary and extensive changes seem as though every writer needed so much help. Thanks also to Bill LaDue, Christine McGowan, Don Yoder, and Barbara Youngblood of Island Press for editorial assistance. But even before the manuscript neared completion, two steadfast friends bolstered my confidence to complete it. Larry Harris challenged me to see beyond the miasma of south Florida's environmental problems and served as my mentor when I returned to academia. Herb Kale inspired my development as a field naturalist during our forays into the waterways of Indian River County long before my interests expanded to include big furry creatures with sharp teeth. He read early drafts and was a constant hedge against lost confidence. Herb was a dear friend, confidant, and champion of Florida's environment—to him this book is dedicated.

Several people were important sounding boards and critical advisers during the development of my thoughts on the Florida panther. Among these respected friends are Jim Schortemeyer, Maurice Hornocker, Kerry Murphy, Kenny Logan, Linda Sweanor, Jim Layne, Mel Sunquist, Sydney Maddock, John Eisenberg, Dave Wesley, Phil Hall, and Tom Hoctor.

It is often said that writing is above all a drain on the families of writers. In our case, the four of us spent innumerable evenings together in a cramped living room, each with some form of homework in hand. My greatest admiration is extended to my wife, Diane, for being the foundation of a healthy family, finishing nursing school despite all the distractions, and tolerating my mental and physical absence from home. No less appreciated are the accomplishments of my children Clifton and Erin who persevered and even blossomed despite their preoccupied parents. No father could be more proud.

INTRODUCTION

There are some animals on earth that have not benefited from the well-intended, hands-on efforts of people. Dusky seaside sparrows, passenger pigeons, and Carolina parakeets all took their last breaths in captivity after humans had destroyed their homes. At the other extreme, coyotes, European starlings, and nine-banded armadillos have all thrived despite locally intense efforts to eradicate them. Like the three extinct birds just mentioned, Florida panthers have gained little by human efforts to manage individuals and manipulate population genetics. The panther's problem, like that of grizzly bears, cheetahs, California condors, and black-footed ferrets, is space. The survival of these species will ultimately have little to do with the manipulation of animals in the wild or captivity. It will have everything to do with the way their landscapes are managed.

The Florida panther is not endangered—at least not by any of its much-publicized enemies. The last panther may well succumb to a poacher's bullet or a vehicle's bumper. Or it may die of an acute bacterial infection. Perhaps like Martha, the world's last passenger pigeon, the last panther will be a lonely captive acting as ambassador for a subspecies that was once spread across a boundless landscape. If the Florida panther is reduced to the point where an individual determines the fate of the population, it will be mute testimony to the worth of today's recovery efforts. For now, the panther is holding up its end of the bargain. It is up to us to let it demonstrate a resiliency hampered only by limited space.

The story of the Florida panther's management in recent history is a classic example of what happens when people treat only the symptoms of a much larger problem. In fact, the inconsistencies in treatments have been so great that it is a wonder the panther still survives. Diseases, parasites, highways, hurricanes, inbreeding, and heavy metals have all been cited as immediate threats to the panther's existence. Yet none of these problems has impaired the panther's ability to live and reproduce where there is suitable habitat. Strategies have come and gone for dealing with these problems: protocols for mercury-poisoned cats, airlifts for the injured, inoculations for kittens, and captive breeding to save them all. Today, five years after ten kittens were removed from the wild, these same cats, now adults, sit behind fences with no plans to breed them, no plans to release them, and little prospect for increasing the wild population's chances for survival. Instead, cougars from Texas have been brought to the steamy forests of south Florida as a quick fix to a complex problem, and soon the Florida panther will be a different animal than the one that has survived all the other attempts to rescue it. It may survive these efforts as well, but then again, it may not.

This book offers a glimpse of an incredibly adaptable animal that has weathered a century's worth of attempts to eradicate it as well as two decades of misdirected efforts to save it. And it truly can offer just a glimpse. Even those of us who know the panther best got only fleeting views of the animals. Once every two years we might see each panther in a tree and then drugged for an hour on the ground where we filled in the missing details of its life by deciphering the scars, dislocations, and tattered radio collars that told stories of rugged lives—lives mostly hidden by dense forests and undetected by our technology. How much more we might have learned if our work had involved *visible* animals! There were no drawn-out bouts of observation where we could record the behavior of panthers killing their prey or the intricacies of panther family life. Because of its habits, the panther is truly a phantom to human observers. But we made the most of our infrequent opportunities. Capturing panthers was the highlight of our work.

It took years of sacrifice by many agency staff to collect enough information to correct the pessimistic image that prevails among scientists and the public. This animal is not a biological lost cause. Yet the panther is still cloaked in a cloud of misconception. For panther recovery to advance beyond our present state of ignorance, the story of its ecology and distribution must be told to a wider audience. The purpose of this book is to paint an accurate picture, not only of the panther, but also of some of the events that have driven a decades-long debate.

One of the key occurrences at the outset of panther research was the accidental death in 1983 of Female 3, an old resident of the Fakahatchee Strand that never woke up from anesthesia. At the time, fieldwork was conducted solely by houndsman Roy McBride, Florida Game Commission project leader Chris Belden, and a handful of other biologists. Even though the unfortunate panther was likely too old to breed, the loss of this cat caused shock waves that led to Chris's removal from the project, the transfer of a new project leader to Naples, Florida, and the addition of a veterinarian to all future panther captures. My involvement in panther research would come only after the new project leader, John Roboski, had endured a frustrating two years that culminated in his abandoning the project and Florida altogether. Constantly handcuffed by administrative decisions made hundreds of miles away, John felt powerless to direct a field project that demanded a fair amount of autonomy. Thus the first five years of panther research were conducted under anything but stable conditions.

At the moment of my first panther capture in January 1986, I never imagined I would be writing anything more than the technical papers expected of a research biologist. A book about panthers, written from my perspective, became thinkable only after nearly a decade of working closely with five remarkable people on this equally remarkable project. We were all employees of the Florida Game and Fresh Water Fish Commission (GFC) panther research team—a collection of people with different backgrounds, different responsibilities, and differing opinions about the panther's history, status, and recovery prospects. Our field activities were closely scrutinized, not only by the state agency that employed us, but by three other agencies that had an interest in the Florida panther. In 1986 these four agencies would form a group known as the Florida Panther Interagency Committee, an organization loosely modeled after the Grizzly Bear Interagency Committee and intended to coordinate the activities of those involved in panther recovery. The U.S. Fish and Wildlife Service (FWS) had developed a plan to save the subspecies and in fact had been the major funder of research for more than a decade. The state's Department of Natural Resources (DNR—now the Department of Environmental Protection) and the National Park Service (NPS) had similar interests as stewards of large tracts of publicly owned south Florida wilderness.

During the first few years of research, the land managed by DNR and NPS was thought to be the heart of panther range. Thus, the Fakahatchee Strand State Preserve, Everglades National Park, and the Big Cypress National Preserve were considered by all agency staff as the keys to the panther's future. Certainly, these were not the only managers of large chunks of south Florida wildlife habitat: the South Florida Water Management District controlled most of what was left of the historical Everglades, the Seminole and Micosukee Indians inhabited large reservations at the edge of the Big Cypress Swamp, and private landowners held a patchwork of panther habitat that equaled the acreage managed by government agencies. None of these groups had an official voice on the Florida Panther Interagency Committee, however, or a say in the direction of panther research and management.

Our work was meant to be the foundation for the panther's recovery. Only a decade earlier, everything known about the panther had been compiled in a mimeographed collection of papers presented at the 1976 Florida Panther Conference, a meeting jointly sponsored by the Florida Game and Fresh Water Fish Commission and the Florida Audubon Society. The meeting was called specifically to pull together everything known about the panther and decide if there was any point in formalizing efforts to study it. Although there was more speculation than facts, spirits were high and a vision for saving the animal emerged. The plans were simple. A radiotelemetry study and eventually a captive breeding program would be needed to save an animal that had not yet revealed its breeding status. For all we knew in 1976, the panther was already extinct. By the time I got involved in the project in 1985 (five years after the inauguration of GFC's fieldwork), nine panthers had been captured and my two predecessors had gone through the emotional trauma caused by frequent panther deaths and an unsupportive administration. Yes, panthers had been proved to exist in the wilds of south Florida. But the handful that had been captured suggested a grim future. Several of my colleagues advised me not to join this troubled project. They cited the political nature of the work, dishonesty among supervisors, severe working conditions, and unpredictable co-workers. Many agency administrators talked openly about panthers as a lost cause. It seemed that for every new capture, two panthers died of vehicle collision or some other calamity. We were just bearing witness, it seemed, to an extinction process.

Had this book been on my mind during our first capture, the concept would have been shaken by what I saw in the tree—the very creature that had created the stereotype of the unhealthy panther. This was not the dangerous predator that struck fear into the hearts of early settlers. Panther 8 was old. She seemed infirm. As she balanced there on the laurel oak limb, her natural tawny coloration and a mottling of muddy swamp water blended sunken features with the tree bark of her background. Number 8 typified everything I had heard about the Florida panther: senile, anemic, parasite-infested, and an ever-present target for Collier County motorists. With only two female panthers (Numbers 8 and 9) wearing radio collars as of January 1986, there was little hope that the panther project had much of a future. The lone radio-collared male, Number 7, had been a highway victim the previous fall. Male 1 was hit by a car in 1983, Male 2 was killed by another panther in 1984, Male 4 was hit by a car in 1985, and Female 5 and Male 6 had died of unknown causes shortly after their captures in 1982. Even after the shuffling of personnel and the addition of a veterinarian following Female 3's sudden death, the study sample continued to decline.

Not only had I inherited little to work with, but field activities had all but ceased by early 1985—and to make matters worse, the unsupervised tracking of radio-collared panthers by a GFC wildlife officer/pilot had led to the disappearance of all our location maps. The result was the loss of nearly a full year's worth of priceless movement data on the handful of cats that were believed to constitute the population at the time. I felt as though we were starting all over again. Nonetheless, there was a constant demand from state and federal agency administrators, the public, and all forms of media for the latest news of Florida's rarest mammal—which led, of course, to the broadcasting of preliminary information and the publishing of unsubstantiated reports. The result was the remarkably enduring image of the panther

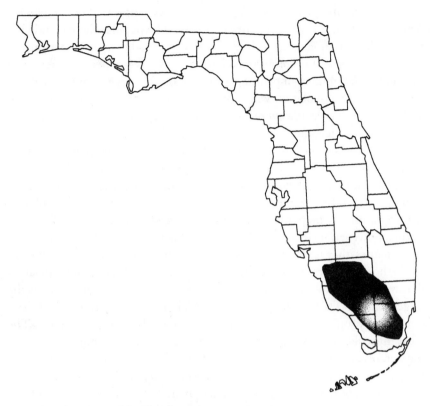

Known breeding range of the Florida panther

as a decrepit, disease-ridden cougar competing with hookworms, ticks, and mosquitoes for its very own life-giving blood.

Reports on the condition of a few radio-collared panthers became the basis for evaluating the entire population—then estimated at less than thirty—while the slow and arduous task of year-round data collection had only just begun. In the eight years that followed my first capture, our Naples research staff authored or coauthored over twenty peer-reviewed papers describing panther ecology and I compiled a technical bulletin summarizing the fifteen years of fieldwork conducted by the Florida Game and Fresh Water Fish Commission. As we began to see that panthers were in fact healthy, long-lived, and reproducing, the field team wrote reports, presented data, and talked with the media. During this span of time my coworkers and I participated in dozens of television documentaries, news programs, and radio broadcasts, and our work was covered by countless newspapers and magazines. Despite all this coverage, or perhaps because of it, the picture of panthers that emerged was as fragmented as the landscape they inhabited. And often it was a highly inaccurate picture. The popular media tended to seize on the sensational; the technical journals tended to take the personality out of our prose. Attempting to convince agency personnel, reporters, and the public that individual panthers were not on the verge of nutritional collapse—that the population was in fact doing more than holding its own—took most of the years covered by this book. Many remain unconvinced, however, and inaccurate images are cherished even today.

Compounding our communications problem was the Game and Fish Commission's insistence that we never use names to identify our study animals. I would never have considered referring to a panther by name in a technical paper. But for the sake of making our message clear to the public, a certain familiarity helps to convey what the Florida panther is. Our administrators, however, were above all concerned about their agency's image. The use of names, they feared, would portray their fieldworkers as too emotional.

In the early years, we made a habit of anointing study animals with names that were linked to some unique physical trait or special location. Thus the largest panther ever captured in Florida, Male 17, became "Jumbo." His frequent and proportionally huge mate, Female 18, became "Jumbette." The most muscular and exquisite panther, Male 20, became "Animal"—he was the stereotype of grace, beauty, and raw feline power. An oft-mated pair, Female 11 and Male 12, were the "Bear Island Female" and "Bear Island Male" simply because that is where we caught them. Occasionally we poked fun with the naming. "Big Al," Male 13, was not all that big, but the GFC's assistant executive director, Allan Egbert, happened to be touring the Big Cypress by helicopter on the morning of Number 13's capture and dropped in to watch the spectacle. Dr. Egbert was so impressed by the event that we named the new cat after him. Big Al was run over by a truck less than two years later.

The occasional use of proper names for our otherwise numbered and sequenced study animals came to an abrupt and official end following the recurring abandonment of Female 23 by her mother and sister, when both kittens were radio-collared in an unprecedented double capture. Number 23 tolerated several bouts of captivity before she finally was repatriated to the wild. In the meantime, however, she had endeared herself to the project veterinarian and some of the National Park Service staff during her rehabilitation, and as a result she became "Little Orphan Annie." No doubt this name was too anthropomorphic, but it emphasized the dangers of capturing entire panther families, especially when more than one kitten was present. (Given her small size, this animal may well have been destined for an early death without our intervention.) An appendix at the back of the book gives the vital statistics of all the panthers we studied between 1981 and 1994.

*   *   *

After fifteen years of research, those of us most intimate with the Florida panther are convinced that all of its problems can be traced to landscape management issues—primarily the conversion of forests to farms and urban development. But these are not issues that generate much popular support. It is far easier to sensationalize the drama of roadkills, disease, and starvation—none of which has really affected the panther's current status. This book represents my attempt to give an objective account of the Florida panther through the eyes of a field biologist and correct the public's biased image of this animal. But the panther landscape is not only the plants, animals, and soils of a watery wilderness. It is also the politics of natural resources that are used by agencies far removed from, and out of touch with, south Florida. And so, to comprehend what a panther really is, we must understand some of the forces behind the push to change it into something else.

The book is a loose chronology of personal experiences that defined how I came to know the Florida panther. These accounts frame detailed descriptions of natural history and behavior that emphasize the differences and similarities between Florida panthers and their close relatives in the western United States, which are usually considered no more than legal game. The book also reflects the evolution of the panther project, a story in which biology was gradually overwhelmed by politics. I do not refrain from criticizing some of the actions of individuals and agencies, but I have tried to balance the critiques with alternative solutions that would return some of the dynamics and optimism that once drove field activities. This kind of drive and interest will be needed if the panther is to survive much beyond the turn of the century.

THE
FLORIDA
PANTHER

GETTING OUR FEET WET

I

13 January 1986 ■ I saw my first living panther on this day in one of the remotest corners of the Fakahatchee Strand. This part of southwest Florida was an ankle-to-waist-deep second-growth cypress swamp into which I was leading our capture crew, a motley collection of full-time and part-time employees of the Florida Game and Fresh Water Fish Commission (GFC), toward Panther 8, an old female with an unremarkable history as a study animal. Other than the grogginess that accompanies an alarm-assisted awakening before daylight, the six of us showed little emotion. The scheduled recapture was just another day of work that promised to shed little light on the status of the population. She was one of only two study animals—remnants of a fitful five-year study that had concluded the endangered Florida panther was a lost cause.

Behind my facade, however, I was feeling anxiety from my inexperience mixed with the suppressed excitement of seeing one of the rarest animals in North America as I did my best to avoid the deep holes and cypress knees that could plunge me yet deeper into the icy January waters. While Roy McBride, our lanky houndsman from west Texas, led his sloshing dogs toward Panther 8's signal, I listened to the receiver's beeping as Darrell Land, a biologist fresh out of graduate school, lost a sneaker to a particularly deep gator hole. Melody Roelke, our husky and energetic veterinarian, chattered with animation to a guest journalist. Biologists Jayde Roof and Walt McCown relentlessly added to a visiting writer's fear of swamps and large, slithering, sharp-toothed reptiles. The fruit of our labor and the hounds' short chase was this thin and grizzled panther. As we readied our safety equipment, she stared at us balefully from 20 feet above in a spreading laurel oak. The capture was uneventful, Female 8's recovery from anesthesia was slow, and none of us looked forward to retracing the swampy mile and a half to our vehicles. The obstacle-strewn route was even more hazardous to cross now under a moonless and star-studded sky. Muffled curses, splashes, and frequent thuds evolved into purple bruises the next day.

If the capture of Female 8 had been a letdown in expectations, Male 10, captured two days later, was just the opposite. As the first kitten to be captured in six years of fieldwork, this five-and-a-half-month-old panther represented a glimmer of hope for the future. Number 10 was the product of the pairing of Male 7 and Female 9, both residents of the Fakahatchee Strand and a future housing development known as the southern Golden Gate Estates. These two adults had been chased up the same tree about a year before, apparently interrupted in the act of mating. Jayde Roof still bears scars as evidence of this unusual event: Panther 9 clung to his backside as she slid, partly drugged, out of the tree. In contrast to his parents', Number 10's capture was uneventful—except that, because of Jayde's sensitivity to poison ivy, I had the dubious pleasure of donning the climbing spikes and extracting the drugged kitten from his refuge in a spindly cypress. With my heart pounding from anxiety and exertion, I unceremoniously pitched the 34-pound panther into the net, then watched as the attention of my coworkers switched from the top of the tree to the capture below. Left unwatched to descend through the tangle of poison ivy, I was thankful for a fairly easy climb and a lightweight panther. I had neither the experience nor, certainly, the strength of Jayde, our usual climber (Figure 1.1). In the years to come we would watch Jayde on innumerable occasions, wedged precariously atop a swaying tree, holding an adult panther by one hand, arm extended for what seemed like hours, before dropping it into our cushion or lowering it slowly down by rope. Many of our successful captures were due solely to Jayde's skill in the most suspenseful aspect of our work.

For the next hour Melody examined the kitten, took blood samples, and monitored his recovery. I fumbled with metric tapes, calipers, and the first radio collar ever attached to a juvenile panther (Figure 1.2). The collar was big enough for an adult panther, and I worried that its size might impede his travels. But since his mother was doing the work for the family, perhaps it would not slow him down. Besides, within a few months a shipment of smaller collars would arrive. Melody noted that Number 10 had pale gum tissue and a dull, rough coat, signs of anemia that could be caused by parasite infestation. Although I had little to compare him with other than a dead panther and old Female 8, Kitten 10 looked pretty good to me and so my field notes reflected the subtle contradiction between our two evaluations. We could only guess at his age—most of the previous year's telemetry data had been lost or simply were never collected while there was no field supervision on the project. So exactly when Panther 9 and her first kitten emerged from

Figure 1.1
Jayde often performed acrobatics while extracting panthers from tall trees.

Figure 1.2
The scene of a panther capture complete with a range of equipment and "guests."

their wooded den was only a guess. Our estimated age for the still-slightly-spotted cat was just less than six months.

His mother remained close by during the workup, circling the capture site until we left. Because of the surrounding swamp, we put Number 10 in a specially designed tent built with coarse-mesh nylon to prevent him from stumbling into the water and drowning before the anesthesia wore off. Leaving the door of the tent slightly open, we left him to be reclaimed by his mother. The next morning the tent was empty and the family was reunited: two radio collars, transmitting on different frequencies, emanated from the same spot in a distant corner of the swamp. Our intrusion into the family life of these two panthers had resulted in only a temporary separation, and we looked forward to the wealth of information this mother/son pair would ultimately provide.

Our accomplishment that day was a significant expansion of the work initiated more than a decade previously by a World Wildlife Fund–sponsored survey by Roy McBride and U.S. Fish and Wildlife Service biologist Ron Nowak. Roy's treeing of an old adult female in Glades County back in 1973 became one of the compelling events that stimulated a wave of new research. They reported:

> One panther was located and live-captured by McBride. Sign of this animal was found on the morning of 10 February near Gator Slough, about five miles west and one mile north of Lakeport, Glades County. The pack of eight dogs picked up the trail of the panther and quickly began active pursuit. After a chase of about 20 minutes, the panther climbed a cypress and was injected with a tranquilizer drug fired from a dart gun. The animal then came down from the cypress, and after a brief chase was treed again in an oak where it was overcome by the effects of the drug. McBride brought the animal out of the tree, and on examination found it to be a nine or ten year old female that had apparently never given birth to young. It appeared to be in poor condition and was infested with ticks. The panther soon recovered from the tranquilizer and was released. Observation of this animal confirmed reports of panthers in the area of the Lykes Brothers Ranch west of Lake Okeechobee.

It took another eight years before Chris Belden initiated the formal investigations that we were continuing. Belden started by logging reports of panthers originating from Florida and attempting to determine their veracity—in line with a 1978 recommendation by the Florida Panther Recovery Team, a group of biologists and others appointed by the U.S. Fish and

Wildlife Service to develop a strategy, known as a "recovery plan," to save the panther. This team of experts was one of the tangible products of the 1976 panther conference. They developed the first written recovery plan that served as the outline for the first decade of fieldwork. As in many studies of wildlife of unknown status—Big Foot and the Loch Ness Monster included—the first task is to quantify the creature's basic distribution. It was remarkable how the widespread sightings and reports of panther sign, especially tracks, generally turned up near human population centers, with a noticeable peak in observations around rush hour (when most people are out and about but most panthers are not). Most reports turned out to be cases of mistaken identity. Belden soon learned that authentic reports came only from south Florida, particularly Collier County. Further, this was the only area from which specimens of actual panthers had been collected (as the result of roadkills) or living animals had been seen in recent times.

Panthers were conveniently rationalized as extinct in Florida by most agency personnel before 1973. Little had been written about them since David Newell's documentation of a 1935 hunting expedition (eight panthers killed) and a brief account of vocalizations and behavior by herpetologist Ross Allen in the *Journal of Mammalogy* in 1950. Jim Bob Tinsley wrote an informative popular account of the Florida panther based on historical notes and anecdotes in a 1970 vignette, but literally nothing was known about this animal's biology. Many of the reports Tinsley cited reflected Floridians' fears of panthers and their ignorance about wildlife in general during the late 1800s and early 1900s. Thus the panther treed in 1973 deserves great accolades. Not only did she demonstrate the panther's existence, but she also represented a bridge between the era of persecution and the era of conservation. In previous years, most of the panthers treed in Florida were no doubt dispatched quickly with lead.

16 January 1986 ■ With the capture of Panther 10 behind us we turned our attention to the Bear Island unit of the Big Cypress National Preserve. Of the 1.5 million acres administered by the National Park Service in south Florida, the 38,000 acres in the Bear Island unit were clearly superior in terms of soil fertility, abundance of forested uplands, and prey density. Deer and hogs were plentiful here, and I hoped to take advantage of this situation by radio-collaring the panthers that made it their home. It was at this time that I began learning the basics of panther sign identification—a frustrating educational process. None of my college training in wildlife at Ohio State or the University of Florida offered the basics of mountain lion sign identification. Like my previous work on Florida bird communities and on Florida black bears, this would be on-the-job training.

Hunts for uncollared panthers were sometimes as brief as a single day when luck ran high and tracking conditions were excellent. More often than not, however, the capture of a new study animal, if it was not the dependent kitten of a collared female, took weeks of reading blurred hieroglyphics left in mud and sand and piecing together the suspicions bit by bit. Our crew would split into groups—Roy McBride with his hounds, the rest of us on swamp buggies or on foot (Figure 1.3). The dogs, oblivious of the distinctive patterns left on the ground by panther feet, cast about for the odors left behind by the cats as they brushed against vegetation or left their scent in a track. As these odors were detected, the dogs would follow them like a rope until the line of scent disappeared or a panther was treed. Sometimes the dogs would retrace an entire evening of a panther's travels, reading the messages left on palmetto fronds, wild coffee leaves, or any of the hundreds of species of tropical plants that are found here. It often took hours to re-create the panther's path on these scavenger hunts, and we commonly covered many miles in a morning. The actual chase of a panther, however, usually lasted just minutes.

Figure 1.3
Handmade swamp buggies were the primary mode of field transportation.

For the next week, although the hounds were unable to detect scent trails fresh enough to follow, we saw abundant panther sign—especially the tracks and scrapes of an adult female. A scrape is the result of a panther scratching up a mound of soil or leaves over urine or feces. These are used as a means of communication among panthers. Adult males leave scrapes throughout the year, but females scrape only when they are in estrus. These scent markers have several functions: they are apparently critical in the spacing of adults, they help males locate estrous females, and they help maintain social structure in the population.

20 January 1986 ■ While accompanied by Department of Natural Resources biologist Ken Alvarez, I came upon the tracks and scrape of an adult female panther along an old Bear Island logging road known as Gillette Tram. The signs were fresh enough to have been made the same day or the night before. A dense canopy of laurel oaks, red maple, cabbage palms, and cypress had helped preserve the fresh appearance of the tracks, but it was too late in the day for the dogs to follow the cat's trail. Red-eyed vireos, great-crested flycatchers, and other resident and wintering songbirds provided a musical backdrop during our mile-long hike into this mixed swamp. And the pungent, crawfish-filled feces of river otters, the rancid scat of a bobcat, the odorless pond apple droppings of a black bear, and the scattered pellets of white-tailed deer reminded us of the other creatures that also depended upon this forest. We were standing in the heart of the only place east of the Mississippi River where this combination of terrestrial carnivore species—panther, bobcat, and black bear—still live.

21 January 1986 ■ Roy and his hounds were out before dawn and hunted in the vicinity of yesterday's sign. The rest of us scouted for new sign or waited along Alligator Alley to intercept any dogs that might follow a scent trail across this busy highway. Despite the debatable notion that vehicles are the single most important cause of panther mortality, there was no debating with Roy the odds of his dogs making their way safely across, especially during the peak of tourist season. Any risk of losing one of his highly trained hounds was too great.

As the morning wore on, the temperature rose to the upper seventies, the wind picked up, and conditions worsened for panther hunting. The protective, jungle-like canopy over Gillette Tram, however, gave the hounds additional time to scour the area around yesterday's fresh sign. Because both wind and evaporation were minimized in this verdant setting, sign might very well last all day. At noon Roy's voice crackled over the radio. He had chased our uncollared female up a tree. Walt, Jayde, Darrell, Melody, and I drove as close to the cat as possible and then hurried on foot to the tree. Standing on a laurel oak limb that was partly hidden behind the fronds of a 20-foot-tall cabbage palm was the healthiest, sleekest-looking panther we had ever seen. Somewhat dazzled, we stood in awe of the tawny, obviously healthy, and well-fed young adult female, now crouched on a horizontal branch. Her attention was focused primarily on the hounds barking below. We were but annoyances—probably no more and no less significant than the occasional pesky mosquito.

Despite the elation of adding an important panther to our study, this capture was notable for its unsettling revelations about the inner workings of the team. The group I was responsible for directing suddenly became a bigger challenge than the research that had brought us all together. With a full two weeks of project supervision under my belt, I felt comfortable with the routine and decided that Female 11 would provide a good opportunity for me to expand my experience. But my decision to exercise what I thought was a right if not an obligation of my job—that is, pull the trigger of the gun that projects the drug-laden dart into the panther—upset a long-standing social tradition of which I was flat ignorant. The instructions I had received from my Tallahassee supervisor, Bureau of Research Chief Tom Logan, said nothing about such intricacies of personnel management and division of labor. These details would have to be learned on the job. It would be up to me to develop the team atmosphere necessary for safe captures even though the standard operating procedures, if any, were a mystery to me. Any number of disasters—ranging from injured people to dead panthers—could result from a disorganized capture effort. As much as Melody was always effusive and ever willing to talk about medical management and other details, Walt, Jayde, and Roy tended toward silence. Perhaps they just assumed that the proper protocol was obvious. Perhaps they were just waiting to see how their new boss would turn out. Only later would I learn that even though the dart hit Female 11 perfectly and the workup of the cat proceeded smoothly, I had unknowingly undermined my relationship with Roy, the resident marksman. Not only that but 21 January was Melody's birthday, and, to everyone's surprise, Walt had hidden a decorated cake in his swamp buggy for the occasion. The celebration was short-lived, however, and subdued. Roy left unusually quickly to tend to his dogs, no doubt dreading future captures with the pushy young project leader. Darrell was off stowing equipment. Jayde, for reasons known only to him, refused to partake in Walt's gesture of friendship to our veterinarian. By that evening, as we prepared for sleep at the Ochopee field station, I felt distinctly uneasy about the project's future.

At the beginning of my supervision of the panther project, I learned more from my staff and coworkers in the field than from written sources. Previous to his annual forays to Florida, Roy McBride was one of the first to study mountain lions with radiotelemetry. He received his master's degree for this work from Sul Ross State University in 1976 following many years as an animal damage control specialist for the federal government in Texas. He has probably legally killed more "panthers" in the western United States than have existed in Florida over the last hundred years, but there is no one better at catching them. And his insight into individual panther life histories was often uncanny. On several occasions after the capture of a new study animal, the weathered, stetsoned Texan would announce some facet of the panther's life that would take radiotelemetry a year to uncover. I often imagined that Roy's dedication to the panther project was his way of balancing his career as a wildlife professional—using the same tools and skills to both kill and rescue North America's most secretive cat.

Walt McCown was a product of north Georgia and Alabama, and indeed he often put on the act of being the stereotypical Deep South redneck. Walt did his graduate work at Auburn University before moving to Naples and the panther project. He was one of the developers of our famous panther capture cushion and was responsible, too, for conducting a deer physiology study. Our largest team member, Walt was impressive for his appearance— always with a machete in hand at captures—but his sharp wit and unexpected sophistication dispelled the Billy-Bob first impression. There were few involved in panther fieldwork who did not experience one of Walt's frequent pranks—like the radio collar planted near my house and attached to a stuffed Teddy Bear.

Jayde Roof was the one among us who actually grew up in Florida. Although he was not of country origin, he was surely at home in the woods. Most people have a fairly poor sense of direction, but Jayde had an unerring sense of place despite his Orlando upbringing. Never in the eight years I worked with him was he directionally impaired—his internal compass took us unerringly to and from the field without a misstep. Jayde was the project workhorse: climbing trees to capture drugged panthers, assisting Roy with his dogs, extracting stuck and broken vehicles from the mire, and flying in rough weather. Although he was reluctant to share his opinions, I suspect his silence was due to his false impression that greater intellect accompanied advanced degrees.

I met Darrell Land while he was still a graduate student at the University of Florida and I was working on black bears for the Game Commission in Gainesville. As I needed someone to fill a temporary position on the panther project, Darrell became the only person on the project that I had a hand in hiring. Perhaps this is why he seemed extraordinarily devoted to me—a blessing that continues to this day. Darrell's talents gradually became more obvious and more important with time, and eventually I was able to obtain permanent status for him. He also had a cast-iron disposition when flying, and probably has logged more hours tracking panthers from aircraft than anyone alive. Darrell promised me in 1986 that it would not be long before he returned to his North Carolina home. But among my five original coworkers, he and Roy are the only ones still part of the south Florida panther project.

Melody Roelke preceded everyone but Roy on the project and had the weight of tenure to back up her strong views on panther biology. Although there are those who claim that the Florida Game and Fresh Water Fish Commission was home to more male chauvinist Neanderthals than were found in the public at large, this did not seem to influence the inner workings of our team. Melody's terrific energy and outgoing personality earned her the nickname "Turbo-Vet"—she ate, drank, and dreamed panthers. Because she felt she had been hired as the antidote to panther capture mortalities such as Number 3's accidental death, she must have suffered the anxiety of these events more than anyone. The possibility of a panther death loomed at every capture. Experience with cheetahs at an Oregon animal park helped her qualify as the panther veterinarian, a position that did not become permanent until several years after it was created. Job security notwithstanding, she worked tirelessly at saving the panther and was very vocal in stating her opinions. Melody was as tough a fieldworker as anyone else on the project. In fact, she frequently carried an overweight pack that contained enough chemicals and equipment to supply several MASH units. Even after she had a full-time assistant, each of their packs seemed to grow. Certainly they were heavier than any single pack the rest of us carried.

As for me, my travels began in Fairbanks, Alaska, where my father wrapped up his army career before returning to Cincinnati. I was never that fond of the cold, so it seemed only natural to continue my migration south when my good friend Herb Kale suggested I enroll in graduate school. My master's thesis dealt with the impacts of phosphate mining on birds, but my wildlife interests were broad enough that I was hired in 1980 as a furbearer biologist for the Florida Game and Fresh Water Fish Commission in Gainesville. I still don't understand why I was recruited so vigorously for the vacant panther research supervisor position by agency administrators. There must have been many trained carnivore biologists with impressive credentials just waiting for a job like this one. Having seen the frustration in Chris Belden and my predecessor, John Roboski, I initially turned down the promotion and pay raise hoping for better things in Gainesville. But finally I gave in—after all, there were only a few years left in the project before we were all transferred elsewhere or assigned to some other work. John Roboski became so frustrated that he left Florida permanently for a forest wildlife job in Alabama. Chris Belden was reassigned to the GFC research office in Gainesville where he would slowly recover from his emotional roller coaster and eventually get back into panther research by studying experimental Texas cougars in north Florida. My own frustration would peak in 1994— the same year I reentered graduate school to earn a Ph.D. on the work that would become such a large part of my life for more than a decade. Depending on one's perspective, I was either tenacious enough or foolish enough to last four times longer than any previous panther research leader.

27 January 1986 ▪ After a weekend break we returned to Bear Island to search for more uncollared panthers. A cold front had swept into south Florida bringing blustery weather that favored the panther and hindered the dogs' effectiveness. Nonetheless, tracking conditions were ideal. We soon discovered prints of an adult male and an adult female, apparently traveling together, in the southwest corner of Bear Island. These were the first adult male tracks I had seen in south Florida since a visit to Corkscrew Swamp Sanctuary in 1981 to examine black bear habitat. Any panther track is large and impressive—even a kitten track is larger than an adult bobcat's—but next to the dainty footprint of the female, the male's track was gigantic. A plaster cast of this tom's front footprint revealed a pad width of nearly 2.5 inches with the characteristic two lobes in front and three in back. The four toes were arranged asymmetrically around the pad with a lead toe (corresponding to our middle finger) extending beyond the others. The fifth toe, or dewclaw, rides high on the carpal joint, the panther's "wrist," and does not make contact with the ground. Because the panthers were walking casually, their retracted claws left no impressions in the black mud. Although we spotted no panthers, the tracks gave us encouragement and afforded a new prospect for our slowly expanding study.

28 January 1986 ▪ Working in the subtropics of south Florida dictates field clothes that are lightweight and quick-drying—especially after a good soaking in a pop-ash slough or cypress swamp. On this particular day, however, all the protective gear in my possession would have been inadequate to ward off the bone-chilling cold delivered by this unexpectedly severe arctic front. Walt was home sick, and Darrell was alone searching for sign on a three-wheeled all-terrain cycle (ATC). This left Melody and me on one buggy, and Roy, Jayde, and the hounds on the other. Given the troubled atmosphere of last week's capture, this arrangement of people, I hoped, would reduce the likelihood for conflict. While Jayde was busy following Roy and the hounds through sawgrass marshes, wet prairies, pine flatwoods, and hardwood hammocks, Melody and I searched for panther sign along some nearby buggy trails. A stocking cap and unlined parka were hardly effective as the winds picked up and the temperature dropped on this clear sunny morning.

We had seen numerous tracks in the drying mud—opossum, raccoon, white-tailed deer, Wilson's snipe, great blue heron, and hispid cotton rat—but no new panther tracks. By 10 A.M. we had caught up to the other buggy where Roy was busy retrieving his hounds. All but one had been loaded into the specially built boxes mounted on the rear of the vehicles. The day's hunt, it seemed, was over.

The wayward hound, Sluggo, was not only a veteran of many west Texas mountain lion captures, but had also taken part in the bulk of the panther work accomplished so far in south Florida. Annoyed that Sluggo had not returned with the other dogs, Roy went trudging off to find him. Because each dog wore a radio collar similar to the ones we used on panthers, it was easy to locate them even when they were too far away to hear their barking. As I was finishing off the second in a package of two frosted cinnamon pop tarts, shivering between each bite, the buggy radio came to life. Roy was telling us to get to him in a hurry. But our short trip was interrupted when my buggy engine sputtered and died at the edge of a hammock and sawgrass marsh. After two weeks of scouring the Fakahatchee Strand and Bear Island for panthers, the buggy's 30-gallon gas tank was empty. We radioed Jayde who met us quickly with the other buggy to collect Melody, our gear, and me. By now we had forgotten the cold. After all, only one thing could have prompted Roy's unexpected summons. Sluggo, despite the wind and without the assistance of his fellows, had trailed, chased, and treed the maker of the male tracks found the day before.

The panther in this spreading live oak glowered at us from behind clusters of bromeliads and the menacing vines of poison ivy that entwined its branches. Because the tree was not very tall and the cat was only about 15 feet up, we opted not to use the crash bag (the portable cushion developed by Walt and John Roboski to break the falls of panthers from tall trees). With only four of us at the tree, we were all glad to have a low tree to work with. Even so, we knew he would fall heavily into our hand-held net. As I put the dart gun together I fretted over the shooting protocol and relived last week's incident. With the rifle and syringe ready to go, I handed them to Roy and asked if he had a good, unobstructed shot at the panther. He accepted the equipment and my unspoken apology, thus easing the anxiety I had unwittingly caused the week before. This restarted the process of building a solid professional relationship—and, for now, allowed us to proceed with the capture of male Panther 12.

The uncollared panther shifted only slightly when the dart hit the muscle mass of his large hind leg. In five minutes he began to lose balance and slid with little resistance into our net. He was an impressive beast—well muscled, sleek coat, a few well-healed facial scars, and callused, well-worn feet that we would find characteristic of adult males (Figure 1.4). A notch in Male 12's left ear suggested aggressive encounters—perhaps with a struggling boar or a dispute with another panther over breeding rights. Both ears received the tattoo "12" and a radio collar was fitted to the Bear Island male's 18-inch neck. Roy, with dour wisdom in his voice, remarked that Panther 12 would teach us a lot about panther ecology. We all wondered where this 122-pound male would lead us in the future. Would he turn up as another highway death statistic, succumb to diseases and parasites, or overturn the accepted hypothesis of panther decrepitude? Regardless of the answers he would provide, my initial image of the Florida panther as disease-ridden and malnourished was brightened by the captures of Panthers 10, 11, and 12.

The early afternoon sun had warmed the air into the lower forties, but with the excitement of the capture behind us we found ourselves redonning

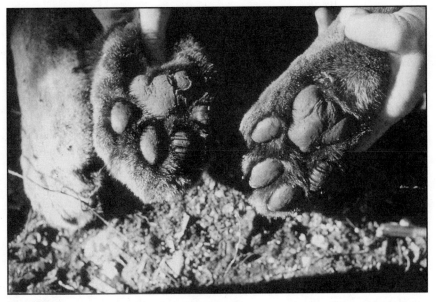

Figure 1.4
The leathery, callused feet of Panther 12 were typical of wide-ranging resident adult males.

jackets and hats and I wished for a pair of gloves. Number 12 received additional drugs to keep him immobilized for a new biomedical procedure, electroejaculation, that added nearly an hour to the cat's processing time. Melody's interest in panther biology had recently expanded beyond simply safeguarding anesthesia to include the documentation of male reproductive parameters. Even before adequate data had been collected to make generalizations on this subject, poor sperm quality and quantity were accepted as characteristics of this subpopulation. Undoubtedly, Melody would have liked to harvest female gametes as well, but we had no simple method to do this in the field. Our estimates of female reproductive health would hinge on our ability to observe kittens or their sign. Melody believed that the panther population was old and exhibited little if any reproduction. Given this preliminary conclusion, based on the captures of a half dozen or so panthers, she thought the problem might lie with the reproductive performance of males.

The procedure began by sliding a stainless steel, lubricated probe into the rectum of the panther. The probe, attached to a 12-volt power regulator, delivered varying levels of electric stimulation to the cat's prostate until drops of semen appeared at the end of his penis. During the peaks of stimulation, Male 12 became rigid while involuntarily extending his toes and dagger-sharp claws until the electric surge subsided. This process was repeated perhaps two dozen times. The products of the panther's stimulation were pearly drops. These we placed on a glass slide with cover slip and examined them under a small field microscope. After several minutes of searching, I was able to find a few slowly jittering sperm cells seeking an egg to fertilize. The low density of spermatozoa in Male 12—and in all other panthers examined in successive years—became cause for alarm and yet further evidence of the Florida panther's bleak future.

My library work, however, suggested that male panthers do not consistently exhibit low levels of fertility. Perhaps panthers were not so strange after all. In fact, Dale Towell and his other wildlife agency coworkers in Oregon wrote that male cougars in their area exhibited seasonal variation in production of spermatozoa. When I brought their manuscript to our veterinarian's attention, Melody questioned its scientific merit. But the biology made great sense to me and suggested a similar cycle might apply to panthers in Florida. Just as sperm production is stimulated by testosterone, so too is aggressive behavior. While seasonality may have been the pattern detected in Oregon cougars, testosterone and sperm production may ultimately depend on *female* sexual cycles. In other words, perhaps male fertility is driven not by seasonal changes in the environment but by the hormone levels of resident females. As we would learn in years to come, male aggression was a common phenomenon when a sexually receptive female was courted by more than

one male. After all, what evolutionary advantage would there be for a resident male panther to be *constantly* on the brink of a fight? Our telemetry data
indicated that chance encounters between two adult males in the absence of
females happened rarely and never resulted in mortality. This may explain
the apparent contradiction we consistently observed in our adult males—apparently low fertility combined with remarkable success at impregnating
their mates. The only females we studied over a fourteen-year period that
failed to reproduce were either very old (Female 8), too young (Female 41),
or suffering from some physiological abnormality (Females 18 and 38). The
other twelve females we studied during this time raised a minimum of forty-
eight kittens out of twenty-four litters.

With the electroejaculation over and all the other procedures complete,
we repacked our gear and began our watch over Male 12's recovery. Darrell
had been picked up by Jayde and rejoined us. After refueling my buggy,
there was little to do but wait and try to stay warm by the little campfire. It
was during this idle time that one of our walkie-talkies, set to scan GFC frequencies, picked up a conversation between two local wildlife officers.
Through fragmented sentences and static we learned that during Sluggo's
solo chase, the space shuttle *Challenger* had disintegrated over Cape
Canaveral. For all of us, this day will always be remembered as one of both
success and tragedy. Our discomfort from the cold—the same cold that contributed to the loss of the entire *Challenger* crew—now seemed trivial. To us
it seemed a bizarre coincidence that both the subject of our work that day
and the terrible human disaster would be commemorated with special
Florida vehicle license tags designed to raise money for charitable purposes.

By midnight, Melody felt comfortable with Male 12's condition. The pig-
gybacking of multiple doses of ketamine to permit electroejaculation had extended the recovery time he needed to metabolize and neutralize the drug.
Six hours after sunset we headed out to the two buggies parked at the edge of
the hammock. A crystal-clear, star-studded sky had descended on Bear Island, sucking out what little heat had accumulated during the day. Upon
reaching our vehicles, we were shocked to find ice coating every metal surface, even the steering wheels. I wished that the adrenaline rush from a new
capture was still in my veins. Perhaps then I would not have been so miser-
able from the cold. Gloveless and otherwise ill prepared for weather more
suitable to my Alaska birthplace, we began our off-road trip back to Turner
River Road. Not until we had emerged from the Big Cypress Swamp at
1 A.M. did Jayde and I pry our stiff and senseless fingers from frozen steering
wheels.

*   *   *

Nearly a month passed before we added another panther to our 1986 sample. Our daily scouring of Bear Island produced no sign of uncollared cats, but it was nonetheless helpful in teaching me the ecology of southwest Florida. Observations of foraging wood storks, a hunting peregrine falcon, a sunning indigo snake, and the nest of a king rail in a burned-over sawgrass marsh served as reminders of the diverse animals sharing this landscape with panthers.

A species seldom seen, but whose presence was frustratingly clear to us every day, was the ubiquitous bobcat. With a body size perhaps only 15 percent of a panther, and with a much smaller home range, bobcats greatly outnumbered their larger cousins. For every panther they pursued in south Florida, Roy's hounds probably chased twenty bobcats. Bobcat chases would often end our hunt for the day because, unlike panthers, these smaller cats had the habit of running from the dogs for hours. Until the bobcat was treed, cornered, or killed by the dogs, we could do little work on panthers. As far as the hounds were concerned, there was no difference between one species that was still legally trapped for its fur and the other that was apparently on the verge of extinction (Figure 1.5). Unfortunately for many bobcats, they did not exhibit the typical treeing behavior of panthers and thus became not only interruptions to fieldwork but mortality statistics. Those bobcats who could not escape the hounds were thoroughly necropsied by Melody and examined for diseases and parasites potentially harmful to panthers. As the panther's closest living relative in North America, bobcats share many of the same parasites and pathogens. And because researchers at Archbold Biological Station had found earlier that disease had decimated a local bobcat population in nearby Highlands County, some feared that a similar fate might await panthers.

Although the bobcat contributed important information, I felt uneasy about this dichotomy in research objectives. Someday we would have our answers about bobcat diseases and would no longer have any justification for their sacrifice. I discussed the subject with Roy. Was there anything we could do to reduce the likelihood of dogs catching bobcats on the ground? I knew this would be neither an easy nor a well-received request. But criticism of our work was already high, and the nature of our research dictated the utmost in environmental sensitivity. Ours was perhaps the most conspicuous wildlife research project in Florida. And there were more than a few outspoken members of the public who were already criticizing the need to harass panthers for study even without knowing that bobcats were being sacrificed in the process. So long as hounds are used to study panthers in the wild in Florida, bobcats will be killed by the hounds. I once considered the use of spring-activated snares to capture panthers, but the potential for serious in-

Figure 1.5
Roy McBride's hounds were a critical element of our efforts to study Florida panthers.

jury or drowning was too great. These traps have worked with tremendous success in our black bear studies, but bears do not depend on speed and agility for gathering food and can tolerate the temporary discomfort that goes with being trapped in a snare. Moreover, bear traps can be more easily set to exclude other species. Traps also worked well for biologists Linda Sweanor and her husband Kenny Logan in a study of cougars at the White Sands Missile Base in New Mexico. In the desert, however, the likelihood of trapping nontarget animals such as bears or bobcats was low and the potential for drowning nonexistent.

From the dogs' perspective, captures of Florida panthers must have been exceedingly frustrating. In Texas and other parts of the world, most of their successful efforts resulted in a lifeless carcass—their reward for a job well done. In Florida, a treed panther, safely lowered to the ground, then surrounded by bothersome researchers, failed to reinforce their correct behavior as effectively as something to sink their teeth into. We always made our drugged panthers available for a good look and a thorough sniff, however, before Roy led them back to our buggies. Bobcats not only provided relevant data. They were the best incentive to ensure that Roy's hardworking dogs maintained their interest in chasing cats. (A few particularly crafty bobcats succeeded in terminating the hunt without being treed or killed by the hounds—either by entering hollow logs or diving into underground holes in the limestone.) Roy understood the dichotomy that our method presented. Over the course of several years, he managed to mold his gang of hounds into a group that was much more tolerant of intervention, yet still effective in hunting panthers. Given the many years of using proven training methods in developing arguably the best cat hounds in the world, this was no easy task. But by the late 1980s, bobcat mortality had dropped to nearly zero.

As part of our greater interest in bobcats in 1986, we began to take advantage of treed bobcats by placing radio collars on them (Figure 1.6). This gave us the first opportunity to compare the spatial relations of the two native cats of the southeastern United States. No interactions were documented between them. And, as expected, bobcat home ranges were comparatively tiny. Males typically used 20 square miles while females used as little as 5 square miles. Bobcats also tended to use less heavily forested areas than panthers. Several of our collared bobcats frequented the edges of active farms or used the interiors of fallow fields. Perhaps this difference helped the two species avoid encounters that would clearly be detrimental to the smaller cats.

Figure 1.6
Bobcats were a distraction to the hounds and occasionally became victims of the chases. This adult male, however, was captured and fit with a radio collar for study.

An adult male was one of the first captures in our unofficial bobcat study. He had experienced severe dehydration during his chase, so we kept him overnight in an air transport crate to monitor him. When Jayde and I checked in on the 20-pound cat around 10 P.M., we expected to find only darkness in the silent lab. Instead we came face to face with an angry, frightened, and unrestrained animal tucked under a necropsy table. Upon escaping his initial confinement in the crate, he found himself hemmed in by the walls of our portable lab and proceeded to destroy all the gloves, tools, tissues, slides, and other sampling equipment stored on the shelves and counters. We quickly retreated to my pickup truck where I prepared a pole syringe, used previously only for bears, to subdue our reluctant subject. Slowly Jayde opened the heavy metal door of the lab and flicked on the lights while I slipped inside with my 9-foot hypodermic. The cat's eyes reflected a fluorescent green from the ceiling lights. He displayed gleaming white teeth then hissed menacingly as I poked the syringe into his thigh. In our haste to escape the wrath of the bobcat, Jayde and I beat a clumsy retreat—the only mishaps being a bump on my head from a prematurely closed door and some additional disorganization to the lab from the bobcat's displeasure.

27 *February 1986* ■ Melody and I were returning the lab terror to his capture site in southwestern Bear Island. The bobcat was wearing a radio collar that would help us understand how Florida's two native cats shared territory. This was the last day of capture planned in the Big Cypress National Preserve for 1986, and I was looking forward to seeing some different areas as well as arranging to move my family from Gainesville to Naples. Most of us were ready for a break in this year's capture work. And after the last few weeks without the discovery of new panther sign, the end of the traditional two months of capture efforts was welcome. The bobcat peered out the open door of the plastic air-travel crate for several minutes—and then bounded through tall grasses and sedges into a dense saw palmetto thicket. As I carried the empty crate back to the swamp buggy, my radio began sputtering and popping with unintelligible fragments of human speech: ". . . ayde-*crackle*-panther-*pop-hiss-crackle*-tree." We had experienced enough garbled communications to know the meaning of this gibberish—especially when what little we could decipher was delivered excitedly by a usually sedate Jayde. The routine now called for me to return an equally clear message. On our end it went something like: "We'll be there as soon as we can." I can only guess what Jayde and Roy heard.

It took us over an hour to return the buggy to my parked truck and trailer, load it, and chain it for transporting the 4000-pound machine to the upper reaches of Bear Island, where Jayde and Roy had planned to hunt this morning. In the meantime, Melody (with a freelance photographer in tow) picked up Darrell at a nearby backwoods airstrip. Darrell had just finished a telemetry flight to locate our five collared panthers. He had heard both ends of our broken radio transmissions clearly and was able to interpret for us. Walt had overheard our radio transmissions, too, because he was counting deer by helicopter with management biologist Jim Schortemeyer and the assistant executive director of the GFC, Dr. Allan Egbert. When I arrived at the tree, I saw ten people standing around the base of a branch-studded live oak before I glimpsed the muscular male panther staring back. This was a stout panther—bulkier than Male 12, but shorter, sleeker, and a few pounds heavier. He exhibited the classic panther traits, cowlick and crooked tail, but he was not scarred and battle-worn like Panther 12. Perhaps he was younger, or just had a knack for avoiding conflict.

Jayde and Roy had taken the hounds down Bear Island Grade, a limerock road maintained by Exxon Corporation to access several oil wells in this unit of the preserve. By midmorning they had picked up the panther's scent and trailed the cat that was now waiting motionless in the tree. Male 13 had cooperated by exercising classic escape behavior that had evolved over millennia and was likely a defense against pack-hunting Pleistocene canids such as gray wolves and hyenas. Number 13's workup went smoothly, and within forty-five minutes the helicopter was lifting its load of dignitaries out of the woods. The rest of us were simply happy to be through with this winter's capture marathon, which had netted half as many new panthers as had been captured in the past five years.

Now we could settle into the routine of tracking our study animals without the distraction of intensive efforts to capture panthers. This study entailed regular flights, every Monday, Wednesday, and Friday, in a Cessna 172 equipped with the same tracking equipment we used on the ground. Because the radio signal carries so poorly through thick vegetation, we needed an airplane if we were to plot the movements and interactions of collared panthers. Ordinarily we could find them all in one or two hours. Rarely did we see them, but our detailed records—plotted on topographic maps, entered on data sheets, and stored in a computer—allowed us to create detailed maps and analyses of their year-round travels. Only very bad weather could ground our flights, and this rarely occurred. Jayde and Darrell always seemed especially resistant to air sickness (Figure 1.7).

Although we captured him on public land, Male 13 was predominantly a creature of private lands. Within days of the collaring, he began a series of movements that introduced us to a landscape where no panther had ever been tracked. The ranches used by Panther 13 represented a distinct contrast to most public lands of south Florida. The better-drained soils and higher

Figure 1.7
A single-engine Cessna 172 was our link to study animals—and a not infrequent source of nausea.

fertility of Hendry County made it a target for agricultural development. The clearing of upland pine forests for cattle pasture had been followed by the spread of vegetable farms, citrus groves, and sugarcane—in all, over 50 percent of its native cover had been converted to agriculture. And this conversion was not yet complete at the time of Male 13's capture. From the air, we easily determined that Male 13's home range was a quiltwork of land uses and cover types. We usually found him in either pine/palmetto forest or oak hammocks. But wetlands were a significant portion of his home range as well. The Okaloacoochee Slough, a meandering north–south wetland system, was the most distinctive feature of Number 13's range—important to local farmers and wildlife alike.

An ancient swamp forest known locally as Sadie Cypress was the natural jewel of the slough, which hydrologically connected the Caloosahatchee River with the Fakahatchee Strand. The centuries-old cypress trees were conspicuous from the air but could be fully appreciated only from ground level. From this vantage point it was clear that their enormity put them in the same age and size class as the famous bald cypress trees of the National Audubon Society's Corkscrew Swamp Sanctuary. The closed canopy of cypress obscured Sadie Cypress's large specimens of red bay, sweet bay, strangler fig, and red maple that formed an equally impressive forest in its own right. During a springtime hike with biologists Jim Schortemeyer and Jim Snyder to examine this primeval remnant, I could easily imagine flocks of Carolina parakeets blending their noisy chatter with the raucous calls of an ivory-billed woodpecker. Many of the large, hollow cypress trees had certainly provided nesting sites for both bird species now long gone, victims of landscape changes that exceeded their abilities to adapt.

THE PANTHER'S LANDSCAPE

2

13 August 1986 ■ One of Panther 13's northernmost locations was on the Hilliard Brothers' Ranch on the east side of the Okaloacoochee Slough near Devil's Garden. The ranch was already devoid of most of its native cover. And of the remaining forested uplands, most were surrounded by clearings created for cattle. Describing the food habits of panthers was still an ongoing phase of work, so we made routine inspections in the vicinity of study animals in order to locate kills, scats, or other signs of their presence. Occasionally we got very close to radio-collared panthers in the process. Number 13's radio signal emanated from an isolated palmetto thicket containing a few scattered slash pines. The 6- to 9-foot-tall palmettos were only marginally difficult for us to force our way through, and within a few minutes Jayde called Darrell and me to his discovery. Number 13, in typical panther fashion, making not a single sound during his exit, had slipped out of the thicket upon our approach. Jayde's find was a black, 60-pound wild hog laying on a platter of pine needles. The panther had probably been busy dining when we disturbed him, for his kill was still ungutted and not yet covered with the usual pile of dirt and leaves. The bright red flesh of the hog's shoulder immediately caught our attention. Blood was still pooled in a few recesses of muscle. The power contained in the jaws of this panther was made obvious by the carcass's missing scapula. Number 13 had unceremoniously ripped the shoulder blade from the hog to expose this meaty part of its anatomy.

As our work expanded to include more panthers inhabiting forested private lands, wild hogs became important in our study of the cat's foraging habits. In fact, the abundance of this food source seemed to account for the better condition of those panthers that incorporated well-forested private ranches within their home ranges. To some degree, the better habitat quality and prey abundance in this part of Hendry County may have allowed the panther to persist despite an increasingly fragmented and denatured landscape. And where native forests still stood, the patterns of human activity were quite different than in the popular wildlife management areas such as Big Cypress National Preserve. Whereas hunting on public land is regulated only by state game laws, on most private land it is regulated by the landowners themselves. Most ranches are leased to groups of fee-paying hunters who not only limit their own numbers and abide by conservative harvest restrictions but have an uncanny ability to detect and deter uninvited guests. In most cases the result usually has been high-quality deer and hog hunts for a limited number of participants—panthers included.

The impact of human disturbance on panthers would become a much more interesting and controversial subject as our sample of collared cats expanded—and as these panthers traveled to different parts of the south Florida landscape. The information we got from these panthers would also become important when it came to dealing with the constant challenges imposed by developers of south Florida's backwoods. For the first time, real data could be cited to modify development plans and direct land conservation efforts. Perhaps our growing database would eventually make a difference in the way south Florida looked at the turn of the century. It had already changed our perception of the panther and its landscape.

*   *   *

Contrary to popular notions, the Everglades have never been a terribly important place for panthers to live. Perhaps the fringe of the Everglades—the migrating "river of grass" made famous by the author Marjory Stoneman Douglas—was some of the best panther habitat in North America, but it is unlikely that many of them ventured into the heart of the 'glades. Undoubtedly the Atlantic Coastal Ridge, which now supports an uninterrupted swath of urbanization from Miami to West Palm Beach, once was some of the most productive forest in Florida, but it has long since disappeared. South Florida is but a small fraction of the panther's original range—now mostly a checkerboard of heavily peopled private property. And within south Florida, most panthers reside in the southwestern corner of the state where forest cover can still be extensive and where sawgrass, sugarcane, and asphalt do not dominate the landscape. Given what little is left of it, it is hard to imagine how extensive Florida panther habitat once was.

Driving Interstate 75 from the Georgia–Tennessee line to its terminus in Miami, a traveler would pass through the historical north–south range of the Florida panther. Another drive from Charleston, South Carolina, to Fort Smith, Arkansas, on Interstate 20 would traverse the panther's historical east–west range. As recently as a hundred years ago, panthers could be found throughout the areas now bisected by these interstate highways: the Ozark Mountains of Arkansas, the southern swamps of Alabama, the hardwood forests of southern Missouri, the Okeefenokee Swamp of Georgia, and the canebrakes of Mississippi. The combined totals for these one-way trips through the panther's original range would cover 1767 miles and take 36.5 hours to complete. A similar highway journey through documented panther habitat today would cover 217 miles and take only 3.3 hours.

The vast area encompassing the southeastern coastal plain includes parts of several regional ecosystems from the southern Appalachians of Georgia and Tennessee, the oak–hickory forests of southern Missouri and north-western Arkansas, the once-great southern pinelands from South Carolina to Louisiana, and the subtropical forests of central and southern Florida. Not only would the astute observer notice a wide variety of biotic communities and topography during this journey, but the influence of human settlement would be obvious. Cotton plantations, cornfields, cattle ranches, industrial tree farms, and the expanding urban centers of the Sun Belt now occupy the best-drained and most fertile lands. The historical range of the panther in the southeastern United States is now a patchwork of farms, cities, ranches, and rural communities connected by a vast network of highways, railroads, canals, and powerline corridors. Within this great web of civilization persist increasingly isolated remnants of a once continuous forest, one that still contains thousands of species of plants and animals. Most of these remnant tracts exist because their trees had no commercial value or their soils were economically undevelopable. A few of the south's forested river bottomlands persist, at least as second-growth forest, because flooding has made agriculture and urban development risky. Rugged terrain keeps at least the ridge tops of the southern Appalachians in forest cover, and here and there larger refugia, including intact hills and valleys, remain as public-owned parks, forestlands, or wildlife management areas. Nonetheless, these fragmented forestlands outside south Florida are not enough to sustain the Florida panther in its historical range. The occupied range in south Florida today accounts for only 5 percent of the area once inhabited by the panther.

The wave of people and agriculture did not reach southwest Florida in a significant fashion until the middle of the twentieth century. Even in early times, Florida was recognized as an exotic retreat with great resort potential. But overabundant water and insect-borne diseases kept Florida's subtropics an inhospitable wilderness that, according to University of Florida botany professor John H. Davis, Jr., supported "a great abundance of the long-legged wading birds, herons, egrets, ibises, many water-fowls, thousands of alligators and snakes, streams teeming with fish, and an abundance of many

Figure 2.1
The popular image of the Florida panther.

of the large mammals such as the deer, bear, cougar, wildcat, raccoon and opossum." After World War II, the elimination of malaria and yellow fever (via mosquito control), as well as the development of modern, affordable air conditioning, transformed an otherwise hostile natural environment into an attractive human milieu. Where well-drained soils occurred, citrus, pastures, and vegetable crops rapidly replaced pine flatwoods and native uplands such as hardwood hammocks. Resort facilities and residences sprang up in coastal areas as mangrove forests and coastal dune forests were reduced in proportion.

The development of the jeep introduced affordable four-wheel-drive vehicles to the public, and soon even the most remote swamps and forests were accessible for human recreation. A dendritic proliferation of paved highways made the settlement of south Florida unstoppable. The largest remaining tract of uninterrupted wilderness in Florida was bisected by State Road 84 (Alligator Alley) when this two-lane highway was dedicated on 11 February 1968. In his 1969 book, *Alligator Alley: Florida's Most Controversial Highway,* August Burghard referred to it as "a delightful surprise . . . a 'must' adventure in highway travel." At the official dedication, State Road Commissioner Jay W. Brown observed: "Alligator Alley crosses one of our nation's last remaining wildernesses, and therefore is deserving of special notice in our tourist-oriented state. For more than three years men have struggled to build Alligator Alley across the Everglades to establish a badly needed transportation link between two of our state's fastest growing areas, the east and west coasts of lower Florida. It is interesting to note that the last time man attempted to build a highway across the Florida Everglades, it took him 13 long years." This was U.S. Highway 41 (the Tamiami Trail), which bisected the same "wilderness," but further to the south, and was completed several decades earlier in 1928.

By 1974, an environmental crisis had been proclaimed by Luther Carter, author of the landmark account of the state's development, *The Florida Experience*. He observed:

> In many places, Florida was becoming dominated by the artifacts of an urban civilization in which nature was too often only grudgingly admitted. No longer the natural paradise described by William Bartram, Florida by the early 1970s was becoming what has been called—with as much truth as exaggeration—"the man-made state." In October 1973, Governor Reubin Askew, addressing a conference he had called on growth and the environment, described the situation in rueful terms: "Let's look around and see what unchecked, unplanned growth has done to Florida. It [threatens] to create megalopolis along the entire length of the east coast and from Jacksonville across central Florida to Tampa Bay and down the south Suncoast. Its waste products have polluted our waterways from one end of the state to the other. . . . It has transformed vast estuarine areas and wetlands into waterfront home sites and canals. It has destroyed beautiful and valuable sand dunes and lined our beaches with hotels and high-rise condominiums . . . resulted in severe water shortages . . . intolerable traffic congestion in many urban areas . . . and threatened [public access to] recreational areas."

At the time Carter's book was published, there was no evidence of a reproducing panther population in Florida. Only a few widely scattered sightings and an occasional roadkill suggested the persistence of a few individuals.

*   *   *

South Florida's climate is considered tropical savanna, with hot, humid summers and mild, dry winters. Occasionally a cold snap like the one we experienced at Panther 12's first capture reminded us that we were not quite in the tropics. While rainfall patterns vary from year to year, 70 percent of the annual average of 53.5 inches falls from May to August. The yearly rhythm of precipitation is sometimes broken in dramatic fashion by tropical storms and hurricanes that usually strike between June and November. These cyclonic low-pressure weather systems originating in warm tropical waters have the potential, not only to cause flooding, but also, due to high winds and surging tides, to drastically alter the succession of south Florida's biotic communities. An average hurricane can dump 6 to 8 inches of rain in a 24-hour period, winds can gust to 200 mph, and storm surges of up to 15 feet above mean sea level have been recorded. Annual precipitation peaks during summer and is primarily the result of afternoon convectional thunderstorms that occur almost every day, usually accompanied by lightning. Historically these summer lightning strikes have been responsible for many fires, although higher water and humidity levels moderated the size and intensity of the burns. Fires during droughts, or human-caused fires in late winter, are usually more extensive and more damaging. Most of south Florida's terrestrial vegetation patterns display the effects of a combination of influences, primarily drainage features, freezes, and fire.

Before the presence of humans in south Florida, fires were almost exclusively restricted to the summer months. But as long as 7000 years ago Native Americans were responsible for many fires as well, and today the descendants of Europeans are the primary cause of offseason burns. Nearly 70 percent of the fires reported in the Big Cypress National Preserve from 1970 to 1977 were caused by arson. Although fire often is perceived as a destructive force requiring suppression and prevention, most of the plants and animals in south Florida are adapted to—and in some cases require—the influence of regular burning in order to survive. In a natural setting, fires are important in recycling nutrients, increasing decomposition rates, stimulating plant growth, triggering the release of seeds, stimulating flowering and fruiting of herbs and shrubs, increasing the amount, availability, and palatability of foods for herbivores, regulating insect populations, and sanitizing plants against disease. No doubt fires set intentionally by land managers have altered the abundance and distribution of native flora and fauna.

Because of south Florida's relatively recent geological age, its soils are poorly developed compared to other parts of the southeast. South Florida is underlain with layers of sedimentary bedrock formed by millions of years of sea-level fluctuations. In most areas, these sediments are covered with materials of recent origin, but where fire has eliminated soils or none have developed, bare limestone may appear as pock-marked pinnacles and solution holes. Organic soils result from the accumulation of plant litter and other vegetative debris, and its depth is directly related to local water conditions. Dry, frequently burned soils, as are found in scrub habitats, are mostly sand, while frequently flooded areas may accumulate organic "dirt" to depths of dozens of feet. When approaching 100 percent organic, these materials are referred to as peat and may support wetland communities ranging from sawgrass marsh to cypress swamp. In general, well-developed and fertile soils support rich plant life, which in turn explains the variety and abundance of wildlife in an area. Thus soils, by virtue of their importance to the plants that prey species feed on, have a powerful influence on panthers as well.

* * *

No other eastern state supports the number of plant species that Florida does—and the southern peninsula is particularly rich, because it supports a flora of both temperate and tropical origin. Among the approximately 1600 species, 61 percent are tropical and 10 percent are endemic to the southern peninsula. South Florida is also home to a large number of rare and endangered plants, especially in the Fakahatchee Strand. As might be expected in an area that lies on the fringe of the tropics, the frequency, duration, and distribution of freezes have had a significant impact on forest composition and tree distribution. This land of intermingling climates has also provided a fertile setting for the establishment of exotic plants. Some of these introductions such as Brazilian pepper, Australian pine, downy rose-myrtle, melaleuca, and water hyacinth have forever altered the south Florida landscape.

Some descriptions of south Florida plant communities are centuries old. And where native vegetation remains, the landscape still bears a resemblance to the forests seen by DeSoto and other early explorers of the New World. To modern eyes, however, south Florida landscapes reflect the changes caused by drainage and other human influences. Except for a few survivors in a few small preserves and their widespread, slowly rotting stumps, none of the huge cypress trees that characterized much of southwest Florida remain. Today, most cypress are thin, straight trees in regenerating forests that are only forty or fifty years old. Or they are the gnarled, stunted hatrack specimens that may be hundreds of years old but were ignored by lumbermen. It is these lesser cypress forests, denuded of their rot-resistant timber or dwarfed by nature, that dominate much of the Big Cypress National Preserve and much of what is considered panther range. Formerly timbered swamps such as the Fakahatchee Strand have grown back as mixed swamps—retaining a dense, foreboding nature but dominated by other tree species such as red maple and cabbage palm (otherwise known as sable palm, the official state tree) that once were towered over by giant cypress. Other swamps are being invaded by slash pine, a predominantly upland tree that is outcompeting cypress where groundwater tables have dropped several feet.

Slash pine is often dominant on well-drained, sandy soils in south Florida. Typically it shares the ground with the fire-loving saw palmetto, the diminutive palm that creates the tropical atmosphere throughout much of Florida (Figure 2.2). From the perspective of most wildlife, saw palmetto is a

Figure 2.2
Pine flatwoods, an important landscape feature for panthers, exist in a mosaic of other habitats.

keystone plant in pine flatwoods. Its saw-toothed, interlocking fronds form impenetrable thickets where young panthers can be raised, where predators can be escaped, and where protection from the elements can be sought. Bears depend on its fleshy heart as a year-round food supply, while nearly all vegetarian and omnivorous animals seek its sticky, black fruit during late summer and early fall. Our team would learn to equate good panther habitat to forests containing well-developed patches of saw palmetto and slash pine (Figure 2.3).

Cypress swamps and pine flatwoods provide panthers with the bulk of forest cover that is available to them. But other woodlands afford important habitat for these large cats as well. Cabbage palms and live oaks form lush islands of vegetation known as hammocks. These areas are frequented by panthers because they produce abundant prey and shady cover. Oak hammocks are often home to a variety of tropical plants ranging from orchids and ferns to bromeliads like Spanish moss and wild pineapple.

We recognized a number of other plant communities that panthers use or traverse. But none were more extensive and fast-growing than those associated with agricultural and other human uses. The area occupied by croplands, pastures, cities, rock mines, roads, and other human uses has increased dramatically over the last 50 years. Hendry County, for example, which forms the southwestern boundary of Lake Okeechobee, was considered 100

Figure 2.3
Their dense foliage and impenetrability make saw palmetto thickets the ideal hiding place
for panthers and many other wildlife species.

percent natural in 1900—yet by 1973 less than half of the county remained
unaltered, with about half in intensive agriculture. This trend is similar for
all of south Florida. One of the fastest-growing agricultural uses of land
within the range of the panther is citrus production (Figure 2.4). With a
southward advancing freezeline, Florida citrus producers have focused conversion pressures on remaining forested habitats. In addition to eliminating
vast areas from occupied panther range (Figure 2.5), agricultural practices in
south Florida have significantly altered the behavior of water on adjacent

Figure 2.4
Citrus groves have replaced productive panther habitat in much of southwest Florida, but forests inhabited by panthers persist in a sea of agriculture.

Figure 2.5
Once they are isolated by extensive agriculture, small forest fragments become nearly useless to panthers and other large carnivores.

forestland. Mostly the region is drier today than before its settlement, whereas other locations may receive more water as the result of ditches, pipes, or other structures. In parts of the Florida Panther National Wildlife Refuge, water depth and flooding frequency in some cypress forests appear to be influenced by field irrigation and draining practices on north-bordering farms. Thus even where wilderness has been set aside as extensive preserves, the influence of people is ever present, even in the remotest swamps.

*   *   *

The Florida panther shares its environment with many other vertebrate species. Some, like the white-tailed deer, bobcat, and black bear, coevolved with the panther in Florida's prehuman wilderness. Deer are an important food for panthers, the bobcat is the panther's closest nondomestic relative, and the black bear is Florida's largest land carnivore, utilizing some of the same foods as panthers. During the 1930s, white-tailed deer herds in south Florida experienced tremendous hunting pressure due to the deer eradication program—a government-sanctioned assault that was triggered by the agricultural community's fear that deer harbored tick fever, a disease that posed a serious threat to the health of their cattle. As a result, an estimated 10,000 deer were killed from 1939 to 1941 in south Florida. Interestingly, it was never demonstrated that wild deer could perpetuate the cattle-fever tick. Another ectoparasite of deer and cattle, the screwworm fly, was eradicated in 1958 and apparently resulted in tremendous increases in deer numbers in parts of south Florida. The impact of these wide fluctuations of prey numbers on panthers is unknown. But the disappearance of deer in the late 1930s likely resulted in at least a temporary shift in panther kill frequency toward the introduced wild hog—which, like another, more recently introduced nonnative, the nine-banded armadillo, plays an important role as panther food. While hogs may have a negative impact on certain plant communities due to their rooting habits, they may have been critical to panther survival during times of thin deer populations.

Among birds, only the wild turkey is large enough to be considered part of the panther's diet, but at least 300 other bird species inhabit the panther's current range. Vertebrates listed as rare or endangered and found within the occupied range of the panther include eleven reptiles and amphibians, forty-eight birds, and fourteen mammals. Of these seventy-three listed species, direct encounters (primarily as food sources for panthers) have been documented only with the black bear and alligator. Invertebrates, though generally small in size, can often be much more conspicuous than vertebrates, and several are known to interact directly with panthers. Over forty species of mosquitoes occur in Florida, and many of them can be encountered as thick clouds during the warm, wet months from May through October. At least eight species of tick occur in south Florida, and one of them (*Ixodes* spp.) is likely responsible for what ultimately develops into the characteristic white flecking found on Florida panthers. Ticks were rarely a bother to us during our many hours in the field. During the wetter months of the year, however, mosquitoes seemed to ignore even the most potent repellents. It was often so hot that our perspiration washed these chemicals away and allowed the pests to feed freely. Due to the general ineffectiveness of commercial repellents in our humid working environment, we usually went without and just tried to keep moving. But walking quickly through south Florida's forests brings its own hazards—such as the sticky traps of the formidable-looking orb web weaver, one of Florida's largest spiders. These colorful arachnids have a propensity to spin their lairs from 3 to 10 feet off the ground, the perfect height to entangle researchers and introduce the panicked, hairy-legged, red-jawed creatures under a collar or up a sleeve. Fortunately for us, orb web weavers would rather eat luna moths than people, and no one was ever bitten. Such cannot be said, however, for *Pseudomyrmex mexicanus,* a dainty, fast-moving ant that lives an arboreal life in the leaves of trees and shrubs. Like its relatives, the wasps, this small insect's sting is worse than its bite. Hardly a day in the field went by without one of us developing a ring of hot, itching welts under sleeves or around the neck.

Several terrestrial vertebrates that once coexisted with the panther are now extinct or have been extirpated during the nineteenth and twentieth centuries. These include the Carolina parakeet, ivory-billed woodpecker, whooping crane, passenger pigeon, plains bison, gray wolf, and red wolf. Among these species, only the wolves were potential competitors of the panther due to similar diets and overlapping habitat needs. Nonetheless, they were part of the highly diverse and colorful biota that existed prior to the European settlement of the southeastern United States.

*   *   *

The five south Florida counties—Broward, Collier, Dade, Hendry, and Lee—cover 4,657,280 acres (7277 square miles) and make up the bulk of the panther's range. Public ownership accounts for 65 percent of this land surface. While such a concentration of public land is equaled nowhere else in the eastern United States, urban and agricultural uses on private lands are on the increase. And the addition of public lands to those already occupied by panthers does not necessarily increase the range of the subspecies—at best it provides protection from development and opportunities for management on remaining panther range that might otherwise disappear. In the meantime, the panther is experiencing continued contraction of its range. In Collier

County, for example, land that was considered urban or rural accounted for just under 24,000 acres in 1973. By 1984 these areas had more than doubled to over 58,000 acres. Similar trends occurred in vegetable crop and citrus acreage as demands for warmer growing conditions stimulated tremendous increases in these land uses. Even though public preserves increased in area, panther range shrank.

As many of south Florida's pristine natural resources were converted to agro-industrial products, efforts were begun to conserve some of the region's unique natural areas. Early acquisitions were the result of private money-raising and corporate generosity. In fact, Henry Ford and his growing automobile manufacturing company attempted to give a portion of the Big Cypress Swamp to the state of Florida in the early 1920s. Although this corporate giant recognized the uniqueness of this area, the state did not. It turned his offer down. More than half a century later, Ford would stir up considerable debate over wilderness development and Florida panthers. This probably would not have happened if the state had accepted Ford's original gift. In Collier County, Collier Seminole State Park was established in 1924, Corkscrew Swamp was set aside as a National Audubon Society sanctuary in 1954, and the Fakahatchee Strand was purchased by the State of Florida in 1974. Public mandate spurred the huge purchase of Big Cypress National Preserve by the Interior Department and the State of Florida in 1978. Although authorized by Congress in 1934, Everglades National Park, located primarily in Dade County, was not established until 1947, after years of debate and boundary adjustments. The Big Cypress Seminole Indian Reservation in southern Hendry County was created in 1938 and currently covers over 42,700 acres. A flurry of land acquisition efforts have taken place more recently. The Florida Panther National Wildlife Refuge, established in 1989, is the first land purchase intended primarily for the panther's benefit. Acquisition efforts have begun in Lee and Collier counties to set aside the Corkscrew Regional Ecosystem Watershed—a convoluted forested wetland needed for regional water conservation that will also benefit panthers and other wildlife. Lee County has led these efforts, but Collier County has failed to live up to its promise of participation. An imaginative land transaction between private Collier County landowners and the Interior Department, however, may result in the addition of over 120,000 acres to Big Cypress National Preserve. And the southern Golden Gate Estates—a 40,000-acre failed housing development adjacent to the Florida Panther National Wildlife Refuge and the Fakahatchee Strand State Preserve—is slowly being purchased from thousands of landowners by the state of Florida.

*　*　*

Hunting has existed in south Florida since the first humans arrived. Native Americans in south Florida, principally the Calusa and Tekesta cultures, depended heavily on white-tailed deer and other wildlife species for food and tools. While the earliest hunters undoubtedly had significant local influence on deer abundance, low human numbers and their primitive weapons likely kept these impacts to a minimum. Their landscape existed as a vast sea of forest broken occasionally by small settlements. But the development of modern firearms, predator control, intensive agriculture, and off-road vehicles forever changed the balance of humans and wildlife. The pastoral hunters of Florida's ancient cultures have been replaced by the mechanized high-tech hunter. A combination of effective weapons and ignorance of wildlife population dynamics led to extinctions of species such as the passenger pigeon and Carolina parakeet. Species such as the black bear and panther were killed by settlers out of fear for life and property, but these carnivores survived because of their secretive natures and their ability to adapt.

Despite the tremendous changes that humans have made in south Florida's diverse ecosystems, the region is still home to the southeast's most wide-ranging mammalian predator: the Florida panther. The variety of plants and animals still present in the current range of the panther attests to their resiliency and resistance to change. Their future, however, is directly related to the speed and magnitude with which people traverse and alter this unique landscape.

AN ELUSIVE IDENTITY

3

Just exactly what is a Florida panther? This question has been asked for more than a hundred years, and there still does not appear to be an answer that satisfies everyone. At about the time that the Florida panther was more the target of fear and firepower than scholarship and science, the nineteenth-century naturalist D. G. Elliott noted that Florida was a popular place for overzealous taxonomists:

> Florida has been a fruitful field for the creation of subspecies, and few opportunities for describing them have been missed, but a number are evidently in a very unsatisfactory state, and require an altogether too large array of witnesses to keep them from falling back into the obscurity from which they have been mistakenly brought.
>
> It is to be hoped, however, that the pendulum has reached the farthest point in its swing towards an extreme radicalism in the recognition of forms, and as it returns to a reasonable equipoise, that a more conservative, and as it appears to many, a more sensible treatment of the often insignificant differences in the appearance of animals may be obtained.

In one of its original scientific descriptions, the Florida panther was considered a unique species (*Felis coryi*) by the naturalist/taxonomist Outram Bangs in 1899, but thirty years later it was reclassified by Edward Nelson and Edward Goldman as just another subspecies of cougar (or puma, or mountain lion). To this day the Florida panther (*Felis concolor coryi*)* remains one of thirty recognized subspecies within a species that enjoys the widest distribution of any mammal in the Western Hemisphere. Considerable variation in size and subtle variation in color reflect the wide range of environmental

*Many taxonomists now prefer to use the genus *Puma* instead of *Felis*.

conditions tolerated by America's native lion. Sea-level scrub forests, Rocky Mountain deserts, South American rain forests, mountains reaching to 13,000 feet—all are home to this adaptable cat. Among living felids, only the common leopard of Africa (and formerly much of Europe) exhibits a comparable distribution and tolerance for different environments. Two common threads link all populations of cougar throughout their range: an abundance of remote, wild terrain and an abundance of large prey to feed on. In Florida the white-tailed deer fills this bill, but cougars elsewhere depend on species ranging from guanacos to mule deer to elk.

Some of the early research conducted by GFC biologist Chris Belden and his associates was concerned with describing the few existing modern panther specimens and comparing them with the few old skins preserved in museum collections. From these investigations emerged the description of characteristics that attempted to standardize the Florida panther. For many years it was believed that for a cougar-looking animal to be a Florida panther it must possess a whorl of hair (a cowlick) on its back, a crook at the tip of its tail (caused by an unusual arrangement of the last three vertebrae), and spots of white hair from the back of the head to the shoulders (Figure 3.1). It just so happens that the area of white flecking occurs on a part of the panther's anatomy that is difficult for it to groom—a part, therefore, where ticks are most likely to survive without being scraped off by sharp claws and teeth. The white flecks, which increase in abundance as panthers age, are almost certainly caused by tissue damage inflicted by tick bites. The other two traits, which are present at birth, are probably genetic in origin. The Florida Museum of Natural History in Gainesville displays a felt-mounted male panther, identified by collection code UF12004, that exhibits a lifelike head containing glass eyes and its own teeth. It was killed near Immokalee on 14 December 1941 by Paul B. Welch Jr. on the Weiland-Welch Ranch in Collier County. The animal possessed a classic 8-inch cowlick and very few white flecks. Although the collection card describes the cat as an adult, the sparse flecking reveals that it was actually a young animal when it was shot. Nonetheless, this fifty-year-old specimen could pass easily for the hide of a Florida panther collected today.

In their classic 1946 text, *The Puma: Mysterious American Cat*, Stanley Young and Edward Goldman described the Florida panther, subspecies *coryi*, in the context of geography and other cougar subspecies (Figure 3.2):

> *Distribution*—Present range: isolated parts of southern Florida and perhaps of northeastern Louisiana; formerly doubtless from eastern Texas or western Louisiana and the lower Mississippi River valley east through the Southeastern States in general, intergrading to the north with cougar, and to the west and northwest with stanleyana and hippolestes. Austroriparian Zone.

Figure 3.1.
Even from a distance, the crook at the end of a panther's tail is evident.

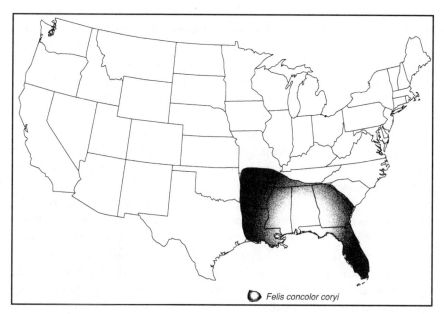

Felis concolor coryi

Figure 3.2.
The original range of the Florida panther stretched from east Texas to South Carolina and included most of the southeastern coastal plain. [After Hall 1981.]

General characters—A medium-sized, dark subspecies, with pelage short and rather stiff; skull with broad, flat frontal region; nasals remarkably broad and high arched or expanded upward. Closely allied to cougar of Pennsylvania but skull distinguished especially by broader, higher, more inflated nasals and more evenly spreading zygomata (zygomata in cougar relatively wider posteriorly, more converging anteriorly). Compared with stanleyana of Texas and hippolestes of central western Wyoming, the color is darker and the skull differs notably from both in the greater general width and upward expansion of the nasals.

Since the total number of Florida panther specimens examined by Young and Goldman was only seventeen, it would be tempting to challenge their conclusions because of the small sample size. How could a scattering of seventeen individuals validate the distribution of a unique subspecies across a seven-state expanse? But no matter how tenuous their descriptions of morphology and range may have been, they were much more important for another reason. In their sections on "distribution" and "general characters," Young and Goldman established that the Florida panther intergraded with several other subspecies, including the eastern cougar, Texas puma, and Rocky Mountain puma. This observation would have significant bearing a half century later as wildlife agencies wrestled with the concept of genetic management of an isolated, remnant population. Without explicitly saying so, Young and Goldman implied that the species *Felis concolor* was distributed without interruption throughout its range. The Florida panther, in other words, was nothing more than one geographic population in a cougar continuum. But would it be safe to conclude that a panther living in a humid, flat, sea-level environment was the same as a mountain lion living 2000 miles away in an arid, mountainous setting?

Compared to other subspecies, the panther is similar in size and weight to the California cougar, but not nearly as large as the higher-altitude-dwelling mountain lion of Nevada. Among all North American cougars, the Florida panther falls toward the bottom of the size/weight scale, but well within the extremes of the species. Adult male panthers average about 120 pounds while adult females average about 80 pounds (Figure 3.3). The smallest adult female we handled in our studies weighed 50 pounds, whereas the largest adult male weighed 154 pounds. The panther's total body length (tail tip to nose) averages 76 inches for females and 82 inches for males. Kittens are born with eyes closed and their soft fur is grayish-brown with dark spots. For the first few weeks of life there is no difference in weight between males and females. (In fact females can be larger than their brothers at this time.) Birth weights of panther kittens are similar to those reported for other puma subspecies,

Figure 3.3
Adult male panthers (*left*) can be twice the weight of females.

though growth patterns vary by location. Panther kittens of both sexes weigh 14 to 18 ounces at birth, and at six months of age males and females are still about equal in size (Figure 3.4). After this age males grow faster and continue to grow for a longer period of time, though females appear to reach their adult weight sooner than males. Whereas it takes males only eight months to double their weight, female panthers require a full year to double theirs. Panther growth rates and weights are similar to the California cougars studied by Rick Hopkins, but their average adult weights are about 13 percent less than the Nevada cougars studied by Allan Ashman and his colleagues. The larger size of cougars living at higher altitudes and farther from the equator is likely related to the thermal benefits of greater body mass in colder climates. The smallest cougar subspecies are found in the equatorial rain forests of South America. Kittens grow quickly: from 25 ounces at eight days, 42 ounces at two weeks, to 65 ounces at three weeks of age. The distinctive spots all but disappear by six months. By the time young panthers are ready to leave the protection of their mothers at no later than eighteen months of age, they have obtained about 70 percent of their adult weight. Growth slows at about two years. Full adult weight is reached at about four years.

As an adult, a panther possesses the classic tools of a solitary feline predator: dagger-sharp, retractable claws and the one-and-a-quarter-inch canine teeth set into short but powerful jaws (Figure 3.5). Although black bears exhibit equally large or larger canine teeth, the longer jaws of this

Figure 3.4
Panther kittens weigh about 1 pound at birth and spend the first two months of their lives in dense vegetation such as saw palmetto thickets.

omnivore contain more variable dentition. Bear teeth are designed for tearing, cutting, and grinding; panther teeth are specialized for making a quick, killing bite and hacking large chunks of flesh from the carcass. Despite possessing the anatomical equipment for killing anything it might encounter, the Florida panther is unlike its western relatives in that it has not been documented as a killer of humans. Although Florida panthers, like cougars elsewhere, occasionally kill livestock, they seem unusually tolerant of people. Anecdotal evidence based on chance encounters between panthers and a variety of observers suggests that in the presence of people an uncornered, unrestrained panther appears unconcerned (unless, of course it is being chased by dogs or otherwise harassed). I know of no evidence that suggests otherwise. While we had few opportunities to test this apparent lack of wariness, no solitary panther ever threatened us, not even if we were alone and tracking it on foot. But Roy McBride has experienced the unnerving situation of a Texas cougar stalking him, and other researchers have come face to face with angry study animals. This has not yet happened in Florida.

Such an absence of wariness has been observed in other species—primarily those living in isolation from predators and without aggressive competitors. The classic examples of this behavior (or lack of it) were observed by Charles Darwin in the Galápagos Islands where the famous finches, iguanas, and tortoises show almost no defensive behavior. Although

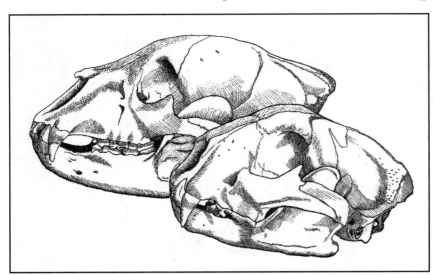

Figure 3.5.
Florida panthers (*right*) have the specialized teeth and skull features of a strict carnivore. Omnivores such as black bears (*left*) possess a more generalized skull shape and tooth arrangement.

this phenomenon seems most prevalent among animals living on oceanic islands, recent studies of biogeography have shown that islandlike populations exist on continental mainlands as well. Given that south Florida is one coastline short of being an island itself, and human settlement is rather recent, is it possible that panthers living in peninsular Florida have lost, through evolutionary time, some degree of their wariness like animals on true islands? How important has this partial isolation been to the making of the Florida panther? As outside genes are introduced to this mainland island population, ongoing experiments with wild panthers in south Florida may demonstrate just how important this trait is.

Tracks, or pug marks, are the most frequently encountered evidence of panther presence (Figure 3.6). These signs allowed us to differentiate males from females and to determine the number of kittens in litters traveling with their mothers. These smaller panther tracks were always identifiable as such because even the smallest kitten track is larger than the biggest adult bobcat track. Although we often encountered otter tracks along trails that panthers frequented, these much smaller tracks usually contained five distinct toe prints and, occasionally, webbing and tail impressions were visible. Dog tracks are often confused with panther tracks, but dog pug marks usually exhibit toe prints with claw marks and the footpad is narrow and less noticeably lobed. Panther footpads have three lobes, and the four asymmetrically arranged toes do not usually show claw marks. Because footpad

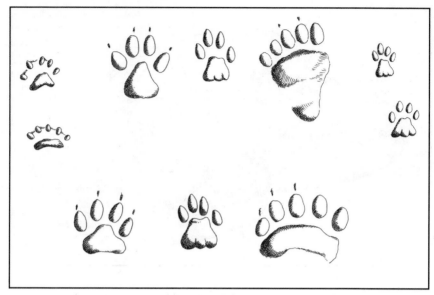

Figure 3.6
Panther tracks are distinguishable by their relatively large heel pad, asymmetrically arranged toes, and the typical absence of claw marks. From left to right (*front feet below*): otter, dog, panther, black bear, bobcat.

dimensions do not vary with the behavior of the animal (whereas running or flexing causes toes to spread), we used this measurement for recording the size of panther tracks. Adult panthers had front footpad widths averaging 2.2 inches and 1.9 inches for males and females, respectively, while the rear pad widths averaged 1.9 inches and 1.7 inches.

* * *

Reported sightings of panthers are legion and widespread. Over the years, those of us associated with panther fieldwork have responded to dozens of calls about panthers in chicken coops, panthers crossing residential avenues, and panthers screaming in backyards. (None of us had ever heard a panther scream, though this behavior has been documented for the species in other parts of North America.) We were always polite to those who reported such sightings, acknowledging the possibility, however slim, that they were true, even though none of them ever resulted in the verification of panther sign in urban or suburban settings.

A surprising number of reports described black panthers, a color phase that has not been confirmed among North American cougar subspecies. I suspect that by virtue of its common name, the Florida panther conjures up for many people the image of a large black cat. This image is so strongly ingrained that the recently formed National Hockey League team, the Florida Panthers, nearly chose a black cat for their emblem and mascot. Among the large, spotted, tropical cats of the genus *Panthera,* black individuals do occasionally turn up. While black leopards and jaguars might be the source of this misconception about the Florida panther, there may be a more local explanation. Florida is unique among the fifty states in being home to the only known populations of native black cats. Resembling more a feline from hell than a bobcat, these animals have turned up infrequently, but periodically, in south central Florida for the past sixty years. (There are no published records before this time.) Their distribution coincides with dark, moist soils associated with the edges of the ancient coastal dune lines of the Lake Wales and Atlantic Coastal Ridges. Several have appeared as road-kills, but the male illustrated in Figure 3.7 was trapped in April 1990 as a chicken-killer and then released several miles away from its point of capture in Polk County, Florida. Curiously, very few reports of panthers, black or otherwise, have come from areas inhabited by North America's only true black cat. Of equal curiosity is the lack of reports of the black bobcats themselves.

Another way of appreciating the identity of the panther is to abandon biology and enter the political realm of the Endangered Species Act of 1973. This forty-five-page document, which was amended by the 100th Congress, defines an endangered species as "any species which is in danger of extinction throughout all or a significant portion of its range other than a species of the Class Insecta determined by the Secretary [of the Interior] to constitute a pest whose protection under the provisions of this Act would present an overwhelming and overriding risk to man." It further defines "species" as "any subspecies of fish or wildlife or plants, and any distinct population segment of any species of vertebrate fish or wildlife which interbreeds when mature." Under these definitions, the panther is eligible for protection under the Endangered Species Act regardless of its genetic makeup. By virtue of its isolation as the last population of *Felis concolor* east of the Mississippi River, the Florida panther represents a "distinct population segment" worthy of recovery.

While this determination should have been an adequate basis for forging ahead on efforts to protect panther habitat, progress was continually diverted by efforts to define a panther genetically. The results of investigations by Melody Roelke and Steven O'Brien of the National Cancer Institute revealed some very interesting aspects of the Florida panther's molecular constitution.

Figure 3.7
The only true black cat native to North America is the melanistic bobcat which is found mostly in southeast and south central Florida.

They found there were sufficient differences between panthers from southeast Florida and panthers living in southwest Florida to suggest that a series of releases of captive-raised individuals in the late 1960s had resulted in the introduction of non-Florida panther genes to Everglades National Park. At least one of the released individuals (presumed to have originated in South

America) is thought to have survived and reproduced—thus explaining the different appearance of the panthers that lived in Everglades National Park between 1986 and 1991. These animals lacked, among other things, the tail kinks and large cowlicks that were so pervasive among panthers living in northwestern Big Cypress Swamp. As a result, questions were raised by agency administrators and members of the public. Did this "hybridized" cougar qualify as an endangered species? Clearly the Florida panther was still an endangered, isolated population of the widespread cougar. Clearly the panther's status became no more or less serious as the result of a series of laboratory tests. Besides, all the genetic analyses were conducted at least twenty years after the supposedly successful releases and more than half a century after the Florida panther subspecies was officially described. Because genetic analyses of Florida panthers were not made *before* the mid-1980s, there is no basis for comparison except with other subspecies. It is impossible to tell how much, if at all, the panther has changed as the result of this presumed introgression.

In 1990, the U.S. Fish and Wildlife Service wrestled with this conceptual problem. In a draft document issued by Regional Director James Pulliam, entitled "Genetic Management, the Hybrid Policy Under the Endangered Species Act, and Species Concepts," the panther's situation was summarized as follows:

> The Florida panther introgression problem is complicated but still appears resolvable. . . . We would be very concerned if introgression now found in some Florida panthers clearly produced consistent morphological or behavioral differences that are distinct from "pure" Florida panthers, other than the loss of characteristics probably derived from extensive inbreeding. Indeed, the loss of probable inbred features through introgression from other mountain lion populations should not be considered detrimental from the perspective of losing genetic integrity of the taxon. Rather, the loss of these characters should be considered essential to maintaining the population. Regardless of whether the Florida panther is considered a population, subspecies, or a species, it is worthy of protection under the Act despite some introgression.

In other words, if panthers were someday to lose their crooked tails and the swirls of hair between their shoulders, this would be just fine so long as the population continued to function normally in its subtropical home. But in the process of losing traits that are assumed to have resulted from inbreeding, would the panther be the same animal that evolved in south Florida's steamy forests? Will the artificial introduction of genetic material from thousands of miles away help the panther more than its long history of adaptation in relative isolation for millennia?

What is a Florida panther? Perhaps it is simply a mountain lion that has adapted to a flat and swampy landscape, isolated from its kin by natural forces and the spread of people throughout North America.

THE NUTS
AND BOLTS
OF TRACKING

4

Most radiotelemetry studies of wildlife start and finish with little fanfare. Study animals are captured, collared, and tracked. Then the data are organized, analyzed, and reported with recommendations to the funding organization. Sometimes, a paper is published. In the many years I was involved in Florida black bear research, we were never confronted with objections to our capture methods despite our use of occasionally injurious methods and the black bear's endearing nature. But none of our actions involving panthers escaped comment from the public. Not a month could go by, it seemed, without a letter to the editor admonishing us for cruelty to panthers—better to let them expire with dignity, we were told, than to prolong an agonizing extinction made worse with humiliating collars, tattoos, and medical exams.

One of the early criticisms charged us with insensitivity in placing radio collars around panther necks. While there was no behavioral or physiological evidence to warrant this concern, ignorance led many to unfounded conclusions. Had more effort been expended in reporting on previous research, these misconceptions would no doubt have been dispelled much earlier. But most government researchers studying wildlife, especially panthers, have been derelict in their responsibility to analyze and publish the data they have collected at great public expense. Our critics included Marjory Stoneman Douglas and many of the followers and fans of this remarkable woman. When asked about panthers, she invariably cited the "documented" panther drownings, changes in their mating behavior, and general harassment with dogs and dart guns. The drownings, we were told, occurred when tree branches snagged collars while the cats were crossing deep water. The futile struggles that ensued ended in the deaths of these unknown victims. I suspect she had been repeatedly provided with erroneous information, perhaps purposefully, by some of her associates. Otherwise there must have been some horrendous goings on before I began work on the panther. My one chance to discuss this issue with her, and assure her that no such calamities had occurred under my watchful eye, occurred at an Everglades Coalition meeting on the River Ranch in the late 1980s. But between the distractions of her many admirers and her focus on shredding the U.S. Army Corps of Engineers, I suspect that my few words had little impact.

Perhaps even worse than agonizing deaths was the thought that data-greedy biologists were more interested in the pursuit of truth than in the preservation of a legacy. Many were sure that the endangered panther would be sacrificed for the thrill of capture and study. According to Florida Panther Recovery Team member Robert Baudy, male panthers always bit the necks of their mates during copulation and the radio collar, a two-inch band of flexible synthetic belting, would either minimize or prevent the crescendo of panther orgasm. It is true that sexually receptive females, captured during or shortly after encounters with males, often exhibited fresh wounds: lacerations and punctures apparently from the claws and teeth of their suitors. But the wounds were usually located well behind the collar. None of our females exhibited scars or wounds around the neck, nor did their radio collars ever appear to have been chewed upon. Nevertheless, the popular platitude was never challenged by the agencies involved in panther research.

If there were widespread doubt about the safety of our research procedures, our ability to document the key elements of the panther's natural history might be compromised. So we were forced to consider this public relations problem. Government agencies are usually quick to react to a negative public image—even when their actions are defensible, even when it means sacrificing important work. Panther research in Everglades National Park was stalled for many years because of National Park Service anxiety about the visitor's reaction to seeing a collared panther. As a result, the extinction process was probably well under way before the first Everglades panther was captured and radio-collared in 1986.

Our newly instrumented study animals, Panthers 11 and 12, provided the beginnings of a compelling story within weeks of their captures—and indeed may have helped us quell the public's growing distrust of our project. Although Male 12 had an unusually long recovery (he remained in a small hardwood hammock for several days), we found him in the company of Female 11 from 4 to 9 February 1986, just one week after his capture. For the next ninety-four days, Female 11 traveled alone throughout Bear Island, showing us each of the corners of her 50-square-mile home range. On 14 May she abruptly ceased her movements, remaining in a dense saw palmetto thicket on the western edge of East Hinson Marsh. For the next two months we spent entire evenings monitoring her activities and documenting her departures and arrivals from her den. Suspecting that she was busy raising her first litter (the product of two radio-collared panthers), we spared no effort in recording this rare look into the previously undocumented process of panther population renewal. Until now there had been little interest in documenting this part of the reproductive process in detail. Past project leaders had monitored instrumented females, but few of these panthers exhibited the obvious pattern of denning (or perhaps the intensity of monitoring was insufficient to detect it). At this early stage in panther research we lacked the state-of-the-art equipment to continuously monitor a female's activity near her den, but we made up for the absence of technology with our round-the-clock presence across the marsh from Female 11 and her newborn kittens.

We established a listening post in a canvas tent on the edge of a hardwood hammock about 200 yards to the east of Female 11's den. Large, spreading live oaks festooned with resurrection fern and bromeliads shared this limestone island with gumbo limbo, myrsine, sugarberry, Spanish stopper, and other tropical trees that shaded our shelter from the searing rays of the early summer sun. Each afternoon the bizarre cacklings of nesting pied-billed grebes greeted us as we set up our equipment to record her activity and movement through the night. The grebes' vocalizations were silenced each evening by darkness—or, more often, by the typical rainstorms created by south Florida's daily cycle of heating, evaporation, and spectacular lightning shows. But following each storm, other animal voices filled the darkness and we were serenaded by tree frogs, pig frogs, narrow-mouthed frogs, chuck-will's-widows, limpkins, and the droning swarms of a dozen species of mosquitoes. In the ensuing thirty-five days we collected over 600 hours of activity and movement data on this one cat and documented the time she spent with her kittens. Although we never saw the den or Number 11 near it, we did not care. Like submariners in pursuit of the enemy, we knew where she was by listening to the sonar-like pings of her radio collar. As in the case of adversarial war games, we had no intention of making our presence known to the new family. Nonetheless, as luck would have it, one of Darrell's late afternoon drives to the listening post, dubbed "Tent City," brought him face-to-face with Female 11. She paused briefly at the edge of a cypress swamp, perhaps incredulous at the appearance of a swamp buggy so long after hunting season, then casually proceeded on her way, disappearing into the tall marsh grass on her evening hunt.

We continued tracking Female 11 long after she and her kittens had left their Bear Island den, and we learned that she successfully raised one male and two female kittens. One of these kittens, Female 19, was captured and radio-collared when she was nine months old. Between 1988 and 1993, Female 19 successfully raised five litters and Female 11 raised at least four more. All of these litters were sired by radio-collared males. Clearly, radio collars did not hinder panther reproduction.

The radio collars also allowed us to administer treatment for some of the injuries the panthers sustained naturally or otherwise. Many situations were handled in the field during routine capture activities—the suturing of a deep abdominal puncture inflicted on Female 18 by an amorous Male 17, for example, or the treatment with Ivermectin of Female 32's mange infection. Some injuries required temporary stays in captivity for observation and rehabilitation. Male 20, a herculean panther specimen, was hit by a pickup truck on a two-lane road on 17 June 1987. An off-duty wildlife officer had collided with the panther about an hour after sunrise. Somehow, injuries notwithstanding, the panther had managed to run off the road and into the nearby woods. Jayde, Walt, Darrell, and I headed immediately to the site to locate the cat and get word to Roy, who was hunting panthers with biologist Sonny Bass in Everglades National Park. Roy and his dogs would be transported by helicopter to the roadside. The dogs would be needed to bay the cat on the ground or, if possible, to tree him so we could evaluate the extent of his injuries. After locating Male 20 in a dense palmetto thicket, I examined the highway while awaiting Roy. A 100-yard-long skid marked the asphalt where the truck's brakes had locked up. The marks were unusual, though, in that one was black and the other brown. Apparently, the panther had not been run over but had taken the impact of the bumper with his chest, and had simply caught a foreleg between a skidding front tire and the road. The oddly colored mark on the highway was composed of skin and the panther's tawny fur.

Jayde, Darrell, and Walt encircled Male 20, still alert, about 200 yards from the highway. While we were anxious to get a look at him, too much disturbance before the hounds arrived could exacerbate his injuries or at least push the cat further from the road. The helicopter landed in a cloud of dust and sand at the edge of a vegetable field along Collier County Highway 858. The dogs were put out immediately to corner their injured quarry. To our surprise, Male 20 got up and ran 150 yards before making an awkward ascent partway up a young slash pine tree. He was clearly in discomfort as he hung atilt from the bottom limb of the tiny tree. A dose of the tranquilizer Ketamine was quickly administered, and within minutes he slid downward into our protective net. His appearance confirmed what the skid marks suggested. The skin and hair were totally abraded from carpal joint (wrist) to elbow on his left front leg, leaving an angry patchwork of exposed muscle, clotted blood, burnt, peeled skin, and dirty clumps of hair.

Miraculously there were no breaks in any of the weight-bearing bones. The nearest facility equipped to deal with this situation was the Miami Zoo, so Melody and I loaded Male 20 into the copter and arrived in the zoo's large parking lot within an hour. The zoo's staff veterinarian, Scott Citino, and a host of helpful curators welcomed the airlifted panther and escorted him into their new surgical facility. About an hour of surgery was required to remove necrotic tissue, clean the wound, and administer topical and systemic antibiotics (Figure 4.1). After a month of rehabilitation at the zoo, Male 20 was returned to his home range where he resumed his customary movements under our watchful eyes.

The technological link between panther and researcher is a rectangular metal canister suspended by rivets from a synthetic fabric collar. The canister is composed of $\frac{1}{16}$-inch brass that measures 2.75 inchs wide by 1.5 inches tall by 2 inches deep. Inside the airtight canister and attached to a system of shock absorbers rests a solid-state circuit board, a tuned quartz radio crystal, a durable lithium battery, and a woven-steel whipcord antenna that extends out of the canister and into the collar fabric. The collar itself is perforated with pairs of $\frac{1}{8}$-inch holes spaced 1 inch apart. This design allows the material to be cinched down with brass and steel hardware to an individual fit and prevents the self-contained unit from coming apart. The entire contraption is then sealed in an amber-colored epoxy that adds more protection to the electronics. Although we prided ourselves on using the best equipment, even top-of-the-line transmitters, like Rolex watches, run low on power and require occasional refurbishing. Every two years or so we would recapture each adult, attach a new collar, and send the old one back to the factory for retooling. In this way, we had electronics that were in service for more than a decade.

All the collars may look the same, but each acts like an independent radio station by transmitting its beeping signal on a unique frequency. The signal of a fully operational transmitter can be detected with a receiver (imagine a glorified radio with tuning dials and volume control) from about a mile away on the ground with a hand-held "H" antenna or as far away as 20 miles if searching for a study animal from 10,000 feet up in an airplane. All panther transmitters are equipped with automatic switches that modulate the interval between beeps. One switch activates a very fast pulse rate if the transmitter remains motionless for more than two hours—usually indicating the death of the cat. Another switch increases the standard sixty beats per minute pulse rate to seventy beats per minute in response to movements of the panther's head. This is how we measure activity.

Most of the radio collars we used were guaranteed to transmit for two years. In later years, advances in battery technology extended that term to

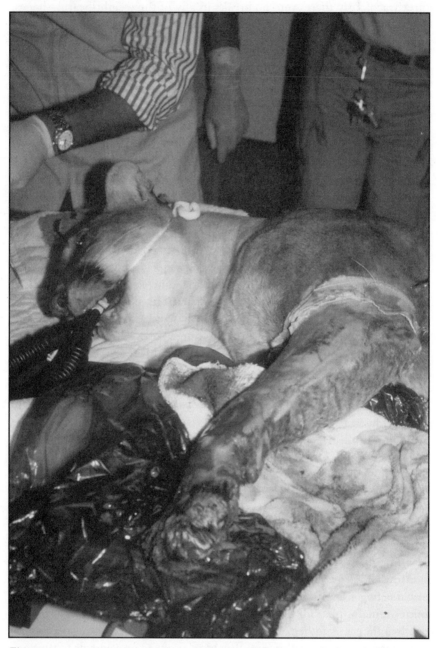

Figure 4.1
Radio collars allowed us to recover injured animals for treatment and return them to the wild—like male Panther 20.

three years. Thus unless something unusual upset the routine, panthers were theoretically immune from capture for at least two years at a time. Our ability to construct a panther's detailed life history depended on hardware that was extraordinarily reliable. But even Rolls Royces and Cadillacs have been known to break down.

26 November 1988 ■ Radiotelemetry equipment can serve a useful purpose in addition to collecting data or monitoring a panther's health—even if the transmitter isn't working. Danny Decker, a technical expert with our supplier of telemetry equipment, Telonics, Inc., helped me in solving a mysterious malfunctioning of a panther transmitter. What at first seemed to be a simple case of equipment failure turned into a learning experience for all of us.

Number 12's transmitter was emitting an unusually slow pulse rate— even slower than the sixty-beats-per-minute head-down signal of a resting panther. For the next two weeks we listened helplessly as his signal got steadily slower until it went out completely on 13 December. Because Roy would not be in Florida until the first of January, we did not have the personnel to assemble a capture crew. But Panther 12 had not exhibited unusual movements outside the home range he had used for the last three years. With consistent effort and a little luck we were bound to cross his path sometime during the upcoming capture season. And so we did—on 15 January 1989. None of us was working that Sunday, but Roy religiously exercised his dogs every day, even on weekends. On this Sunday I was in Sarasota at a meeting with my friend Herb Kale. It was a shock to be handed the phone and hear Walt's voice saying that Panther 12 was awaiting us in a tree. We apologized to our hosts for leaving so rudely, then beat a hasty retreat. If we were lucky, the sun would still be up when we got to the tree.

After miring the swamp buggy twice in soupy mud holes, we found ourselves in a dry cypress forest with the silhouette of a male panther outlined by an orange sunset. Within minutes we were operating by flashlight. Number 12 appeared to be a wreck. Freshly healed scars and a broken toe were new additions to his appearance since we had seen him a year before. He possessed all the key traits of a dominant male panther. The superficial injuries were evidence of defending breeding rights to the females in his home range. Satisfied that he was in better shape than he looked, I turned my attention to the radio collar that had been on the fritz for a month. Nothing was obviously wrong with it. But when we peeled back the protective coating of polymer and exposed the brass radio canister, we uncovered a dent and a $\frac{1}{4}$-inch-diameter hole piercing the stout metal (Figure 4.2). Everyone concluded that

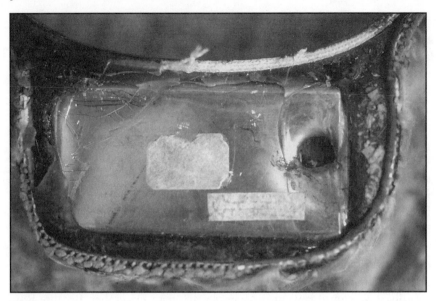

Figure 4.2
A fight with another male panther resulted in a canine puncture that ruined Male 12's radio transmitter.

a bullet was responsible for compromising the airtight environment that contained its valuable electronics.

But a bullet would have left two holes—and likely a dead panther as well. Danny Decker's analysis of the broken transmitter revealed that water had seeped slowly through the epoxy coating and into the canister, causing corrosion and short circuits that ruined transistors and drained the battery. He guessed that months may have passed before enough water seeped through to cause the failure. At this point we looked beyond November 1988 and the start of the strange transmissions for an explanation to the mystery. Our location data revealed that Number 12's last known encounter with another panther had occurred in August 1988. The result of this meeting was the death of Panther 25, a transient male we had captured the previous winter. Panther 25's injuries were not in themselves life threatening. Instead, a severe bacterial infection, probably introduced by Panther 12's claws or teeth, killed him days after their encounter. Number 12's physical appearance at his impromptu capture suggested that he was the loser in the fight. Clinching this scenario was the fact that the upper canines of an archived male panther skull fit perfectly into the dent and hole of the brass canister. It appeared that the radio collar may have deflected a powerful bite by Panther 25 and thus spared Male 12 from a fatal wound.

While these examples dramatize the value of the equipment we used to study panthers, our day-to-day experiences indicated that radio collars were mostly benign intrusions into their lives. The worst that could be said about this technology was that it occasionally interfered with population dynamics by increasing the life spans of a few panthers wearing collars. This seemed to be a very modest sacrifice on the part of the panthers.

*   *   *

We could easily spend more than a month hunting an individual panther (its presence verified by tracks and other sign), but it was the final few minutes that determined the outcome of a capture. By the time a panther was treed, the capture crew often was scattered over a wide area. Using receivers to locate the signals of radio-collared hounds, communicating with unreliable walkie-talkies, homing in on barking, we would make our way to the tree as quickly as terrain and distance allowed. When possible, we drove our swamp buggies close to a spot where we could bring in extra equipment and supplies. Just as often, however, the hounds treed the panther more than a mile from the nearest trail. For these captures, we carried equipment on foot to the tree.

Panthers climb as high as 60 feet to escape the hounds. Their perches range from spreading live oaks to tall, clear-trunked slash pines, cabbage palms, and cypress—three species that present additional safety problems during a capture. Pines and palms can be very tall and without many branches to assist the climber, while cypress are often so studded with branches that even seeing the cat is difficult. The process of firing a dart with immobilizing drugs can trigger several responses in a panther. Some jumped from the tree after being darted—retreating to a dense thicket where the drug took effect and we could safely complete the capture. More often, however, panthers did not jump from the tree after darting—in this case the effects of anesthesia resulted in a cat being lodged in the tree or falling out.

For those that remained aloft, Jayde usually ascended the tree with climbing spikes and a rope. When the panther's drug took hold, Jayde would tie a rope securely behind the forelegs and lower it to the ground. He often contended with breaking branches, large tree girths, slippery bark, and lofty working conditions with few footholds. Retrieving of treed panthers was dangerous, if not just uncomfortable, for all involved. The other variation on tree extraction resulted from the drugged panther's loss of coordination while trying to maintain balance on a horizontal limb. For kittens and most adults treed below about 20 feet, the capture crew held a rope net to catch falling panthers. Panthers falling from above this height exceeded the net's capacity for safety, however, and the strength of our straining hands. In the early years several panthers dropped heavily without a completely broken fall. At least one panther death in the early 1980s may have resulted from such an event. For this reason Walt McCown and John Roboski developed a portable, inflatable cushion that reduced the chances of injury to the drugged panther (Figure 4.3). When inflated, the crash bag is about the size of a trampoline, 10 feet on each side and 3.5 feet high. The green ripstop sailcloth material is shaped by stuffing it with as many heavy-duty trash bags as possible. Inflating a plastic trash bag is a fine art. If you did not develop the proper technique of whirling about while holding the mouth of the bag open to catch what little breeze there was to offer, it had to be filled by lung power. We sealed the bags shut with twine and crammed them into the crash bag which, in turn, we closed with nylon rope. It was often a comical scene enacted under the anxious eyes of many a panther.

Then it is a waiting and guessing game—waiting for a dart to hit its mark and the drug to take effect, guessing the proper placement of the crash bag (if, indeed, the panther falls out). It was an odd, but not surprising, observation that if the cat were to fall it would surely land directly on a person holding the net. This usually meant we had placed the cushion correctly. One of the most astonishing views we witnessed lasted all of three seconds but seemed like an eternity as a drugged Male 26 dropped like a rock from his 60-foot perch in a slash pine to land right in the center of our portable landing pad. In retrospect, I am less amazed by our accuracy in positioning the cushion than by Roy's ability to lob the dart into his thigh (Figure 4.4). Although no one was ever hit by a plummeting panther, neither did every capture go as planned.

We managed an equal number of memorable captures without the need of the crash bag. Number 19's second capture on 9 November 1987 could not be delayed because she had grown considerably since her first capture seven months ago and now needed an expanded collar. It was when she chose to climb to the top of a 40-foot cabbage palm that I asked Roy for his advice in getting her out. Although the lack of branches would have made the climb easy for Jayde, the large cluster of loosely attached palm frond boots and hand ferns beneath the cat rendered her inaccessible. Roy confided that in his native Texas he had routinely cut trees down in order to get cats out. In every case, Roy said, they leapt from the tree long before it hit the ground and then treed again in a more accessible location. This sounded like a workable plan.

Although this species of palm is Florida's state tree, they are on the increase in south Florida, so without hesitation I sent field assistant John Kappes back to the vehicles to collect an ax. In the meantime, the rest of us prepared an elaborate system of belaying lines by tying Jayde's heavy synthetic climbing ropes as far up the trunk of the palm as possible. These, in turn, we fastened to nearby oaks and pines. This arrangement would slow

Figure 4.3
The portable, inflatable crash bag increased the safety of captures for panthers that fell from trees.

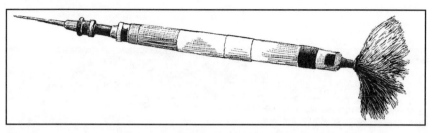

Figure 4.4
An anesthetizing dart—ready to load into the air rifle and fire into a treed panther.

the tree's fall, we reckoned, and give Number 19 more time to negotiate her exit. When John returned out of breath with the ax, Walt, Jayde, and I took turns swinging it against the tree. Number 19 remained still and virtually out of sight throughout the commotion. When the palm began to sway with each strike, those of us holding the ropes tensed for the fall that would bring the crown of the palm into the only clear spot in this large hardwood hammock. We had measured it perfectly. The tree would land exactly where we expected. But our "well-designed" system of rope restraints simply rolled down the trunk of the palm like rubber bands around a pencil. They could do nothing to break the fall of tons of cabbage palm. In muffled horror I watched as the tree rushed toward the earth with no sign of Number 19 nor her flying leap that Roy had predicted. With 5 feet to go she appeared, emerging headfirst among the blurred green of palm fronds, and somersaulted over damp brown leaves and moist sandy soil before beating a hasty retreat to a live oak 200 yards away. Later, as we finished the workup of this healthy young female, I remarked to Roy, "I thought you said panthers always jumped out of the tree well before they hit the ground." Without hesitating he replied, "I didn't say anything about panthers. I said cats!" And indeed he had. He was referring to his many experiences with Texas bobcats! This was the first and last time we used the Texas Ax Method on panthers.

Because the female segment of the population was our window into the world of panther reproduction, the addition of adult females to our study was always viewed as something special. Number 36's first capture on 27 January 1990 began in a low, spreading laurel oak on private lands east of Bear Island. Although she was no more than 15 feet up and we probably could have used only the net, I opted to inflate the crash bag just for added insurance. Number 36 was almost within arm's reach and she looked very large. She clearly was not lactating and had an uncharacteristically straight tail. As she straddled the narrow branch of the laurel oak, her belly, apparently full of deer or hog meat, sagged to one side. Her head appeared oddly small compared to her massive torso. As the bag was being unfurled the panther became agitated, arose, then leapt over our heads and into a nearby cypress swamp. The undrugged cat was immediately pursued by the hounds into hip-deep water where she was quickly surrounded at the base of a rotting cypress stump. A dog would occasionally lunge at the cornered panther who would respond by spitting with gleaming white teeth and exposed claws. The deep water reduced the chances for direct contact as Jayde, Roy, his son Rowdy, and Walt each restrained a dog or two. I waded out of the hip-high water to fetch our stand-in veterinarian, Scott Citino of the Miami Zoo, a loaded syringe, and the dart gun, which I then handed to Roy while keeping an eye on the tensed, angry cat. Busy worrying about the frenzied hounds, the panther hardly noticed the pop of the gun or the impact of the dart in her shoulder. As the drug took effect, Female 36 slowly slumped to the ground and the dogs became silent. We carried her back to dry ground beneath the original capture tree, where the crash bag lay useless. Her apparent size had been no illusion. She weighed in at 108 pounds and 80 inches long—measurements that exceeded Male 37's weight and Male 13's total length.

Now and then we were honored with the presence of a VIP from the Game and Fresh Water Fish Commission's office in Tallahassee. On these occasions there was added pressure to ensure perfection, and so we smothered our ad lib comments and normally instinctive actions. Our guest on 14 March 1991 was the division director, Frank Montalbano. We had scheduled a recapture of Panther 31 in a remote stretch of swamp in the Florida Panther National Wildlife Refuge. Meeting early in the morning, I had sent Jayde and Roy with the dogs to begin the chase while the rest of us hovered about on swamp buggies, staying close enough to hear the dogs but far enough to be out of their way. When we got word from Roy that she had been treed, Walt and I donned heavy packs and began our trek with Frank through the swamp. I had earned a sort of reputation for finding the thickest routes in and out of capture locations, and today was no different. The dogs were close, but a deep, log-strewn swamp was between us and the cat. Ordinarily this just meant getting wet, but I soon began to develop serious concern for Frank and his racing heart, obviously laboring to negotiate the terrain. Walt and I would have had a hard time carrying him out if need be, so we slowed our progress to a crawl while the panther and the rest of the crew waited for us.

Emerging at last from the swamp and arriving at the edge of an open cypress prairie with our exhausted administrator, we could see Female 31 perched nearly 40 feet up in a large cypress. The gnarled branch she rested on was festooned with giant bromeliads, so only her tail and hindquarters were clearly visible. Frank wiped away a tear—of relief? of elation?—as we walked to the base of the tree. There was no discussion as the crash bag was inflated. Number 31's first capture two years ago had ended with an impressive leap from a 20-foot-high slash pine branch and chase into a palmetto thicket. I had little doubt that her second would be very different. The height of the tree would preclude jumping, so it would be up to us to position the bag properly (or for Jayde to work his arboreal magic).

We all looked skyward as the dart found its mark, causing the cat to shift position and send bromeliad fragments into our faces. Looking away was not an option because we might have to slide the bag around the tree if Number 31 changed position. As the rain of debris ceased we became witness to a 92-pound panther sailing free against an azure blue sky. What an incredible sight! Legs outstretched and tail flagging like a rudder. Few animals could take such a risk without the likelihood of serious injury. She landed in a full run and disappeared across a short grass prairie studded with cabbage palms. Because she was drugged and already wore a radio collar, we left the dogs behind and ran after her in case she headed toward water. Luckily Female 31 succumbed to anesthesia in a dry palmetto thicket about 400 yards away. Later, we repacked the unused trash bags for the next "sure thing."

Panthers in southwest Florida were treed in five native trees. Most (58 percent) treed either in live oaks or laurel oaks between 15 and 30 feet off the ground. Panthers may prefer oaks as escape trees because frequent branching, large limbs, and horizontal branches are easier to climb and may provide more concealing cover. Considering all captures in trees, 60 percent ended less than 30 feet up. Slash pine and oaks (upland species) accounted for 67 percent of all capture trees. The high rate of upland species and lower rate of wetland trees chosen by panthers as refuges reflects the proportions typical of panther habitat use. They simply spend most of their time around these upland species. It is interesting, though, that the lowest rate of falling occurred among panthers treed in cypress. These trees, especially of small to moderate size, often had trunks heavily studded with small branches. The greater surface area afforded by cypress branches may have reduced the likelihood of falling. Perhaps the most significant lesson we learned after one hundred tries was that the outcome of panther captures could not be predicted. A safe capture seemed as much a product of our safeguards and technology as it was the result of the panther's toughness. I give most of the credit to the panthers themselves.

A CHANGING
PERSPECTIVE

5

Our captures in the winter of 1986 set the stage for revising what we knew about panther reproduction. For one thing, we began to direct our capture operations to the north and away from the previously accepted center of the population. We were not unlike expectant parents in our anticipation of each new litter. Our experience at Tent City with Female 11 had revealed the comings and goings of a mother panther around her den, while our routine flights kept tabs on her widespread hunting forays. We now attempted the same close monitoring of all females and their kittens.

We had many reasons for wanting to understand panther reproduction. Clearly the future of the subspecies depended on sufficient kitten production to replace the inevitable loss of adults. And based on the last five years of research, it seemed that every panther death represented an irreplaceable subtraction of one from the population's total. Melody targeted hookworm infestation and low genetic diversity as causes of reduced kitten survival, yet we had no hard data to support that view. In our 1989 paper, "Early Maternal Behavior in the Florida Panther," published in *The American Midland Naturalist,* we compared Female 11's 1986 profile of den activity with another panther with kittens, Female 9. Both females were young and in the prime of their lives in 1987. Number 9 was on her second litter, while Number 11 was on her first. Each panther selected dense, concealing vegetation for her den, and for about two months they both took regular forays for as long as thirty-six hours to make kills before returning to their kittens. But most similarities ended here. By the time the kittens were six months old, their mothers gave the impression of being separated by a continent rather than a two-lane highway.

As solitary adults, Females 9 and 11 had home ranges between 40 and 50 square miles. As soon as they gave birth, their home ranges shriveled by

80 percent to less than 10 square miles. The demands of helpless, nursing kittens kept their mothers close to home. This pattern was maintained for nearly two months as the kittens quickly gained size and coordination. Number 9 abandoned her first den after forty-six days and moved her litter about 200 yards to another fern and shrub thicket at the edge of the Fakahatchee Strand. Number 11 maintained a single den in a dense saw palmetto thicket. As Number 9's litter grew, so too did her home range until, at six months of age, the family occupied a larger area than before the kittens were born. Number 11, on the other hand, despite increasing food demands and less experience raising kittens, used an ever smaller area as her litter grew.

The differences in maternal behavior between these two panthers seemed all the more remarkable when we learned that Number 11 had raised three kittens while Number 9 raised only one. Given equal resources (den cover and prey), we assumed that the panther with more kittens would be forced to work harder and travel more widely to find enough food. And yet, by making a series of gradual home range shifts, Number 11 found an area where food seemed unlimited. Number 9 had no other foraging options than those offered by the wetlands-dominated forests south of Alligator Alley and west of the Fakahatchee Strand. Number 11 found a prey-rich private ranch that was dominated by productive upland forests. Number 9 scraped to get by (Figure 5.1).

Although the fates of three of the four kittens produced in these two litters would remain unknown to us, the discovery of kitten tracks with those of Female 9 and the behavior of Roy's hounds (chasing several leads simultaneously) during a recapture of Female 11 suggested that all had survived to independence. Number 9 apparently struggled to raise her single kitten and needed a much larger area to do so. Number 11 could probably have raised a larger litter. Our comparison of Females 9 and 11 offered a compelling account of the inequity of resource distribution in the panther landscape. It was also an important step in demonstrating that parasites and genetics should not be the primary focus of management and recovery.

We spent several years trying to understand one of the resources that probably influenced Female 9's difficulty and Female 11's success: prey. Throughout the cougar's extensive range, one of the prerequisites for permanent occupation is the presence of a large and abundant herbivore to feed on. These prey species range from mule deer and elk in western North America to pacas and European hares in South America. In pre-Spanish Florida the white-tailed deer was the panther's primary food source. Walt McCown's main responsibilities from 1984 to 1990 involved documenting the white-tailed deer population's health and abundance. His findings clearly demonstrated that deer were more abundant and more productive in habi-

Figure 5.1
Female 9 was smaller, had a larger home range, and raised smaller litters than her counterparts living just a few miles away on better-quality habitat.

tats occurring on the more fertile soils and in more upland vegetation (such as Bear Island) than in habitats with generally poor soils dominated by wetlands (such as the Fakahatchee Strand). Number 11 was the first panther to show us that resistance to naturally occurring pathogens was likely enhanced by good nutrition—and could, through high productivity, slow the genetic erosion of a small, isolated population. Number 19 was the only one of Female 11's litter of three that would reveal her secrets to us. (Her brother and sister eluded our capture attempts by seeking refuge on an inaccessible private ranch and dispersing before they could be captured and collared.) But she was important in demonstrating that the widespread claims of poor panther reproduction were simply not accurate.

9 *February 1987* ■ On the day of Number 19's first capture, Female 11 and her uncollared kittens were located in an oak hammock in north central Bear Island. The hammock was less than half a mile from Number 13's 1986 capture site. After approaching on swamp buggies to within a quarter of a mile, Roy and I walked a wide arc to the north of Female 11's radio signal, hoping to intercept the trail laid down the night before by this panther family. At the edge of an isolated cypress dome we sloshed out of 6 inches of claylike marl and onto an old buggy trail. There we saw the clawless cat

tracks, arranged in four sets, confirming that the family had traveled together into Bear Island. The larger tracks headed south with purposeful strides, while the accompanying pug marks of kittens etched random serpentine trails around and through their mother's. For every stride that Female 11 took, the kittens had romped, explored, and skidded twenty.

The chase was short and uneventful and ended with the first female kitten treed in seven years of study. My concern had less to do with getting her out of the oak than with assuring that she was quickly reunited with her family. I relayed this concern to Melody. We should keep our handling as short as possible, I suggested, and avoid administering additional drugs when the cat was on the ground. Unlike the common wildlife anesthetizing drugs of the 1960s and 1970s, ours had no known antagonist (reversal agent). Ketamine hydrochloride just wears off slowly as the animal metabolizes the substance. It is a biochemical relative of PCP—or "angel dust," a fairly common street narcotic. Although the new drugs made wildlife captures safer for people—biologists working alone on a large carnivore such as a grizzly bear ran the risk of accidental self-injection and death because an antidote was required to reverse the drug's effect—this often meant a longer and less safe workup for the animal. While ketamine is extremely safe for both researcher and study animal, recovery time can often run into hours, especially if multiple doses are given.

In the case of Number 19's first capture, I wanted her handling kept to a minimum even though certain biological samples and measurements might have to be sacrificed. At nine months of age, she was probably too young to fend for herself. And abandonment by her mother could have had negative consequences not only for the panther population but also for the continuation of the project. Because Melody did not comment on my suggested low-impact approach, I assumed we were in accord on the impending capture. Darrell, Walt, Jayde, Roy, Toni Ruth (Melody's assistant), Ronnie Bell (a recent university graduate and field assistant), and I were all prepared to physically restrain Number 19 if necessary, attach a radio collar, and get as many samples and measurements as the depth of anesthesia allowed.

Roy's first dart would have hit squarely in the kitten's thigh had a quarter-inch-thick twig not intercepted it—a shot undoubtedly impossible to repeat. The second dart hit Number 19's leg and within ten minutes Jayde was in the tree with panther in hand. Since her 49 pounds eliminated the need for the rope or crash bag, he dropped her the 20 feet to our waiting net. Number 19 was sufficiently still when she hit the net, and there was no need for additional restraint. Nonetheless, as we congratulated ourselves on another smooth capture, Melody plunged the needle of her preloaded syringe into the kitten's thigh. Now there would be plenty of time to obtain the multiple blood samples, nasal swabs, urine collections, and skin biopsies desired for a postcapture tissue analysis. But what had happened to our prearranged plans?

If we did not already recognize the divergent philosophies of the wildlife and veterinary professions, by the end of this day it was quite clear. Veterinary practitioners, by definition, are principally concerned with the welfare of individual animals. The practice of veterinary medicine is taught and conducted, for the most part, in a controlled environment where outside influences are minimized. One of the goals of chemical restraint in veterinary medicine is "good anesthesia"—a state whereby the animal is under complete sedation, exhibiting stable vital signs, and promising a predictable recovery. I suspect that the social bonds in wild family groups receive little attention in veterinary school.

Wildlife professionals, on the other hand, are principally involved in the study and management of populations. A management action is good if it tends to promote the well-being and natural balance of the population. Increasingly, wildlife management involves complex interactions whereby individual species are considered just one of many natural functions of the landscape. In the case of Number 19, she was one part of a complex system— a panther family about which we knew next to nothing. In our only other capture of a panther kitten, Number 10 and his mother reunited immediately after his capture. But he was the only kitten in this litter with fewer (and perhaps stronger) bonds to be broken. There was no way of predicting how siblings would complicate the reunion.

From the veterinary perspective, then, the individual is the central and most controllable element. At Number 19's first capture, I suspect that Melody's habitual faith in controlled anesthesia outweighed, in her mind, the benefits of a speedy workup to reduce the recovery time. But as Number 19's recovery stretched from minutes to hours, the rest of her family continued their travels deeper into Bear Island without her. For several days after the capture, Female 11 and her other kittens were apart from Number 19. Only after the lone kitten and the rest of her family had found their separate ways to a 16-square-mile ranch just north of Bear Island did they reunite. How close we came to causing our first collared female kitten to be abandoned will never be known. The fact that they reunited, however, was enough to reinforce the standard capture protocol with future kittens. Because our capture team operated under a split chain of command (my direct supervisor was in Tallahassee while Melody's was in Gainesville), I had no direct control over the medical management of panthers at their point of capture. Thus lengthy anesthesias and slow recovery times would characterize panther handling for many more years. And this was perfectly acceptable to agency decisionmakers who had added the veterinarian position after the death of that old, postreproductive female in 1983. As handling procedures became more complicated and protracted, I began to wonder if this approach to veterinary medicine might not be a hazard to the subspecies it was designed to protect.

With Number 19 and her siblings safely tucked away under the protection of their mother and with no additional invitations to conduct captures on Collier County ranches, we finished the capture season by hunting for new cats on a large ranch in Hendry County. This expansive 100,000-acre property, owned by Alico, Inc., and chaired by the famous citrus baron, Ben Hill Griffin, was a nearly pristine remnant of wild southwest Florida. Past timber harvest, abundant populations of wild hogs, and grazing cattle had done little to change this productive chunk of the county. The headwaters of the Okaloacoochee Slough wound their way through the ranch among scattered improved pastures, live oak hammocks, and extensive islands of slash pine and saw palmetto. From the air, the slough appeared as a miniature version of the Everglades—with water too deep and widespread to allow for effective draining. The perils of this muddy-bottomed marsh are memorialized by the rusted hulk of an ancient Model A truck, now a permanent monument inhabited by paper wasps and boat-tailed grackles. The truck and its crew had probably made their way across a drought-stricken countryside half a century ago before encountering the mud hole that still claims it.

During our early days of hunting panthers here and familiarizing ourselves with the system of buggy trails, Roy demonstrated one of the hazards common to south Florida's back country. Shortly after he had taken his dogs to hunt for panther sign on an isolated pineland system known as Wild Cow Island, Roy meekly called us on his walkie-talkie for assistance. Jayde and I arrived in my swamp buggy to find Roy's power wagon, ashtray-deep in the slough, not 20 feet from dry land. All we got out of him for explanation was a sheepish grin and "I thought it would make it." I have no doubt that Roy had a momentary flashback to west Texas where he was used to plodding through arid, rocky high country on horseback. When the slough is full, as it was in the winter of 1987, it is terrain better suited to airboats than anything with wheels. But the combination of productive soils, well-drained uplands, and vegetation-filled marshes made the ranch a haven for panther prey including deer and hogs. Not only was there plenty to eat here, but the system of hunting leases on the property, as well as treacherous terrain, kept human activity at a low level.

19 February 1987 ■ Earlier in the week Roy's hounds had trailed a panther for miles through inundated terrain north of Wild Cow Island before the trail literally evaporated. A freshly killed and partly consumed 60-pound hog and a line of fresh tracks were evidence of an adult male panther. The night before this unsuccessful attempt we had camped in an ancient live oak hammock in order to get the most out of the cool weather by being in the field when the sun came up. Barred owls began their evening discourse of hoots and cackles as we grilled steaks, gorged on cheesecake, and finished the evening with a few card games played by the fire. At dawn we were on our way out of camp when Darrell discovered a freshly constructed panther scrape not 50 yards from the nearest tent. Despite our presence, or perhaps because of it, this panther had strolled by our camp, serenaded by the clamorous duet of Walt and Jayde's snoring, and left his calling card with no concern for the consequences.

10 March 1987 ▪ Roy was determined to prevent such a brazen cat from getting the last laugh. We had split into two buggy crews to scour Little Mustang Island and a system of east–west running uplands and marshes paralleling County Road 832. After three hours of panther-sign hunting in the early morning, Jayde, Ronnie, and I had stopped for breakfast at an old hunting camp on stilts. I was hoping that Ronnie had brought along another sample of his mother's home cooking. Two months earlier, after learning that his weekly care package from Fort Meade had arrived, we got a taste of Mrs. Bell's famous German chocolate cake. Immediately we locked Ronnie into a large culvert trap for bears, commandeered his video camera, and recorded him, white knuckles wrapped around iron bars, begging his mother to send another cake. The ransom note and photo were sent next-day air to effect his release. The tape, unfortunately, was seen as simply a joke. So today we found ourselves instead, legs dangling over the edge of the elevated porch, munching pop tarts, sardines, and "wish sandwiches." Our view to the south was a wet prairie surrounded by pine flatwoods and framed from our raised vantage point by Spanish moss, resurrection fern, and twisting live oak branches. It was a placid scene often captured in watercolor and oil paintings hanging in air-conditioned bank lobbies throughout Florida. And, just like these paintings, there was a great egret standing patiently, knee-deep in pickerel weed, waiting for an unwary pig frog to make a fatal movement.

Our repast was interrupted by the crackling of my buggy radio. I wiped sardine oil from my hands and onto my agency-green workpants as we descended the creaking wooden stairs of the camp. Once again we went through the routine of deciphering fragmented sentences to understand the weak broadcast. Walt's buggy radio had died earlier that winter, so he was forced to climb a tree and attempt to communicate with us via a less powerful walkie-talkie. We finished the transmission with questions that were answered only by radio clicks: one for no, two for yes. Two clicks answered my question about the presence of a panther in a tree and we were on our way, hell-bent to the east on our suddenly nimble and speedy two-ton swamp buggy.

The scenery of a wild yet denatured landscape flashed past us as we rolled along at the buggy's top speed—about 30 miles per hour. Our oversized "turf and field" tires, unaccustomed to hard surfaces, threw limestone grit back, up, and into our eyes as we passed cattle pastures, live oak hammocks, drainage ditches, and mile upon mile of barbed wire fence. Baleful range cattle shuffled out of our way or stood chewing contentedly as we rumbled across roadway barriers designed to keep cattle in but let vehicles through. Sunning turtles slid into canals and alligators submerged as our sudden movement disturbed their siestas. Although much of this large ranch was converted to open pasture and plans were under way to introduce sod farming and large citrus operations, significant stretches of forest remained. High prey densities, extensive forest cover, and low levels of human activity allowed panthers to live in a forest that had been fragmented by modern ranching and agricultural practices. The new man-made landscape, more open than the original pine-dominated terrain, was also a boon for species such as Audubon's caracara, loggerhead shrikes, eastern meadowlarks, and coyotes.

Panthers, however, generally shun open, treeless countrysides in favor of deep forests—hunting the confluences of different habitat types and traveling wooded trails. Much of Alico's Hendry County ranch was already off limits to panthers, and this trend was getting worse. Perhaps these large, artificially open landscapes helped explain the absence of female panthers here. Since 1981, no female panther had been captured or monitored on the ranch. Did the predominance of open artificial grasslands (known as "improved pasture") and other agriculture reduce the quality of habitat below a threshold of fitness for females? Or was it this arrangement of habitat features that reduced the level of psychological security required for female panthers to establish home ranges and reproduce? Or had female panthers just not yet recolonized this part of their former range since the days of deer eradication? I hoped they would have time to recolonize before the males abandoned yet another part of their former range.

As we closed the gap on the rest of the crew, our communications improved and within thirty minutes we were parked by Walt's buggy. We could hear the barks of Roy's hounds blasting from a nearby live oak hammock. A hike of 200 yards brought us to the base of a short but widely spreading live oak. Not only did the panther dwarf the branch he was balanced on, but he gave the impression of impending action. His muscles rippled in a way I had never seen in a wild animal (Figure 5.2). He was the picture of raw animal power, the Arnold Schwarzenegger of the panther world.

Figure 5.2
Male 20, an inhabitant of mostly private lands, was one of the most impressive-looking panthers we handled in two decades of research.

Despite looking like a coiled spring, male Panther 20 stayed on the oak limb when the dart hit his thigh. Only a brief snarl betrayed his displeasure. This one, we guessed, was destined for a flying leap and another chase. But within five minutes of the dart's impact, he teetered and fell into our outstretched net and crash bag. With a surge of adrenaline that temporarily counteracted the ketamine hydrochloride, Panther 20 took several short bounds around the crash bag like a gymnast on a trampoline—and then he chose his escape route. A blur of movement included the collapse of Ronnie's tripod-mounted video camera, people scrambling around and over the green deflating cushion, and 148 pounds of teeth and claws hurtling past my head. (I must have been the least intimidating human in the bunch.) Somehow Ronnie, Jayde, Walt, and I were able to converge on the cat before he disappeared into an adjacent palmetto clump. In his drugged state, there was no aggression in this powerful male, but his resolve to escape made even the best handholds tenuous. His slick and shiny coat, the product of an unlimited diet, offered us little help. Our combined 750 pounds of mass created only a slight advantage over the cat, but within five minutes the drugs took full effect so we could release our grips and begin the workup procedure. Male 20 was the largest panther we had captured in seven years of study. Number 17, an older resident, would later tip our scales at 154 pounds at his second capture in 1989, but for now Number 20 held the record at 148. We estimated Male 20's age at three to four years, a young adult male likely in search of his own home range and exclusive rights to breeding females.

By now we had encountered the full array of sex and age classes typical of this species and many other solitary carnivores. And the majority of animals we handled were youthful cats or adults in their prime. The emerging picture of panther social ecology was identical to many populations of mountain lions in the west. The biggest differences had little to do with behavior or biology. The crucial factor was landscape. We were beginning to view the panther simply as a swamp-bound lion with many of the same tendencies exhibited by western cougars. Over the next year, Male 20 would demonstrate to us his tenuous status in this water-dominated wilderness. He was old enough to have survived dispersal from his natal range, but not yet sufficiently assertive (or lucky) to have staked out his own breeding territory.

Of more importance, however, was another finding. Without exception, every new cat we had added to the collared sample since early 1986 was the picture of health. Was this just coincidence? Or was our northward push in search of panthers introducing us to a new segment of the population? It was with a fair amount of anxiety that I reported this information at occasional staff meetings with administrators. After all, it contradicted the first five years of research. Since the recapture of old Female 8, we had captured no senile cats, none that were suffering from anemia, and none that were compromised with starvation. Yet these facts were insufficient to dispel the agencies' notion that the average panther was no more than loose skin over grizzled sinew.

SEX, SPACE, AND PANTHER SOCIETY

6

Among the world's large cats only the African lion is truly social—in fact, its gregarious nature is more typical of canids such as timber wolves and wild dogs. Most felids, including Florida panthers, are essentially solitary predators and the family group, composed of the female and her offspring, forms the basic social unit. How can we account for these differences in sociability? The sparse but even distribution of prey across a forested landscape in south Florida stands in stark contrast to the huge migratory aggregations of prey available to lions across the savannas in East Africa. Differences in terrain, vegetation, and prey undoubtedly help to explain the divergent social systems of these large cats.

Collaring panthers was just the beginning of the arduous process of monitoring the behavior of radio-collared animals. In mountainous terrain, high elevations are helpful because they increase the area that can be monitored—and often several study animals can be monitored simultaneously. Without this advantage, we employed the services of a Cessna 172 and an expert pilot, usually Peter Baranoff, from Naples Air Center. Cruising at altitudes of up to several thousand feet, we flew from panther to panther, covering millions of acres with each flight. When a collared cat was found, Pete would spiral the plane down to about 500 feet where we could identify the vegetation type from which the radio signal emanated. We then recorded its location on a topographic map. Panthers were located by air at least three times each week and opportunistically on the ground when we were engaged in other field activities.

Though flat topography and dense vegetation limited the range of our equipment when we tracked panthers on foot, we did not need to worry about many of the problems associated with tracking study animals in the mountains. Wing surface icing, dangerous downdrafts and updrafts, and unpredictable weather are common conditions in mountain lion country such as Idaho and Montana. In south Florida, airsickness was the most common drawback. Jayde and Darrell had a cast-iron resistance to nausea. They flew, undaunted, into gusting cold fronts and through the unsettled thermals of summer afternoons. Walt, on the other hand, turned green at the thought of flying. I fell somewhere between the extremes: for every two or three good flights, I would have one that sent me into nauseated convulsions that drained my strength and left my head throbbing. Chewing a bologna sandwich and calmly sipping a diet cola as the fits of his passenger's nausea subsided, Pete learned to tolerate what he must have considered a nuisance. But the views were spectacular. And besides, this was the only way to gather enough information to really understand how the panther used its landscape.

At the end of each flight, we would record panther locations on paper forms and enter them into a computer. Later we would use these data to analyze how panthers interacted with the landscape as well as with each other (Figure 6.1). In addition, this information was useful in estimating population size in areas that appeared to be good panther habitat but where telemetry data were lacking. First we would plot a line around all the telemetry locations from the year with the largest number of monitored panthers. Then we would extrapolate this sample density estimate over their known range in southwest Florida. This line, "a concave polygon," eliminated areas that were not used by radio-collared panthers and helped us to avoid overestimating occupied range. We then determined activity patterns by measuring changes in pulse rates emitted by the motion-sensitive transmitters that each panther wore. In later years, we monitored female den activities continuously with an automatic data recording system—a welcome replacement for our twenty-four-hour tent monitoring methods. This state-of-the-art equipment was encased in a weatherproof container, connected to a 12-volt battery and an antenna, and positioned 100 to 200 yards away from the den. We called this system our "biologist-in-a-box." We checked it every forty-eight hours when monitoring dens in order to replace chart paper and exchange batteries.

Estimating the abundance of populations is one of the most basic, yet one of the most difficult, questions facing wildlife researchers. It is one thing to census human populations or animal species that can be readily observed such as migrating wildebeest or wintering waterfowl. It is something else by far to count a population whose members leave little evidence of their presence, are rarely seen, and are especially tricky to capture. Nonetheless, after years of study and slowly building upon the two panthers that remained collared in 1985, we reached a point where additional hunting in those areas accessible to us added no new adult cats to the sample. The first time this oc-

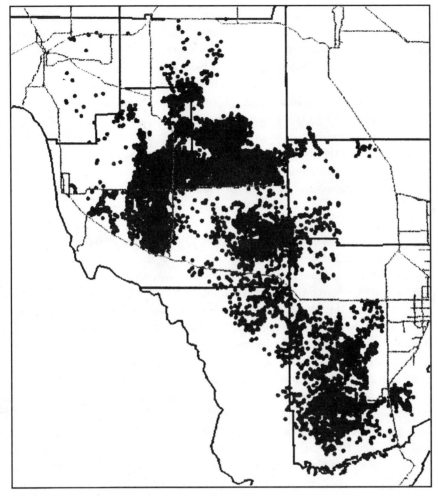

Figure 6.1
Without plotting and analyzing the movements of individual panthers, telemetry locations would appear almost like random specks on a map. Here, several years worth of telemetry locations (unbounded by polygons) suggest some pattern to the landscape that is important to panther distribution: the availability of dense forest.

curred was in 1990. In that year we estimated that the density of panthers north of Alligator Alley and west of State Road 29 was one panther per 42 square miles. This represented the largest number of instrumented panthers monitored at any one time and included four uncollared adult females that were verified to exist by field sign and assumed to be residents. When we extrapolated this figure over panther range in southwest Florida (1,945 square miles)—assuming an even density and similar age and sex composition over time and space—we concluded that in an average year this entire landscape could support forty-six adult panthers (nine resident males, twenty-eight resident females, and nine transient or subadult males). And based on a mean litter size of two kittens and a two-year breeding interval, twenty-eight dependent kittens could be expected each year. The total population in south Florida is likely higher, however, because this estimate excluded the few panthers inhabiting Everglades National Park and the eastern Big Cypress National Preserve, where panther numbers appeared unstable, and ignored Glades and Highlands counties, where access for research on private lands was restricted. A reasonable extrapolation would yield a total southwest Florida population estimate of seventy-four panthers.

* * *

The simplest examination of panther activity consisted of recording pulse-rate changes of panther transmitters monitored during routine telemetry flights. In the few minutes it took to locate each cat, we made a notation of active or inactive. Based on several thousand relocations, panther activity was highest around sunrise, tapered off toward mid-day, then increased again toward sunset. Several factors undoubtedly influenced this pattern, including habitat type, temperature, precipitation, prey behavior, and proximity of other panthers. Most radio locations we collected during the daytime indicated that collared panthers were inactive in dense upland forest cover.

In the few instances that panthers were seen, the cats were usually disturbed while they were sleeping at their midday rest sites. We conducted these ground-tracking exercises to verify habitat conditions seen from the air and to search for panther kills or other signs of their presence. During a cloudy morning flight in January 1991, Darrell saw adult Male 26 in a motionless, prone position—an unusual observation inasmuch as the few panthers we saw from the air were almost always walking and quickly sought cover as the plane circled. Despite his low-level flying, the cat appeared oblivious to the engine noise and remained in the open. This situation was so strange that Darrell began to wonder whether Male 26 was dead or seriously injured. As soon as the dense cloud layer dispersed to reveal a hot mid-morning sun, however, the panther rose and entered a palmetto thicket. Number 26 apparently felt secure in this open situation so long as he was protected by cloud cover. But when the clouds disappeared, he retreated to the dense cover afforded by nearby vegetation.

Darkness seemed to afford panthers an additional element of cover. While we never located radio-collared panthers far from dense forest, tracks and other sign suggested that panthers might travel far from their typical daytime haunts. Observations of tracks in Highlands, Hendry, and Glades counties indicated that panthers occasionally crossed large expanses of cattle pasture to reach other forested tracts or to pursue prey. In Collier County we found tracks leading into large treeless wetlands such as East Hinson Marsh, and Roy's hounds trailed panthers that made extensive movements in other open settings. Compared to the discovery of tracks and other sign in forested settings, however, evidence of panther travel in open country was sparse.

Panthers apparently avoid certain landscape features, both natural and human, regardless of the cover that darkness may provide. During the year immediately after we radio-collared subadult Male 28, his wide-ranging movements covered a three-county area (Figure 6.2). Although he traveled across busy highways, suburban areas, airports, small streams, and railroad beds, Male 28 never crossed the 100-yard-wide Caloosahatchee River. In fact,

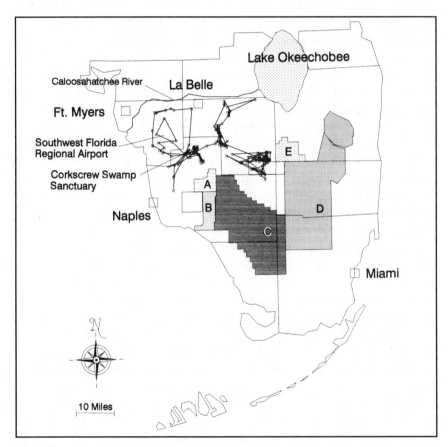

Figure 6.2
Male 28 moved widely through southwest Florida but remained mostly on private land and did not cross the Caloosahatchee River, an apparent movement barrier to most panthers. A: Florida Panther National Wildlife Refuge; B: Fakahatchee Strand State Preserve; C: Big Cypress National Preserve; D: South Florida Water Management areas; E: Rotenberger/Holey Land Wildlife Management areas.

the Caloosahatchee formed the northern boundary of this panther's extensive wanderings. Small watercourses did not impede the movements of panthers, although they avoided swimming when possible. Weirs and bridges
spanning such canals as the L-28, Faka-Union, and those along State Road
29, and Alligator Alley have all been regularly used by Panthers 26, 9, 11, and
12. These crossing points varied from grassy culverts and concrete bridges to
narrow steel "catwalks" for workers' access to water control structures.

*   *   *

From the perspective of panther evolution, highways are a significant but
novel landscape feature in south Florida. Although highway collisions are a
well-known cause of panther mortality, less is known about the influence of
roads on panther behavior. Sign and telemetry indicated that panthers
readily traveled unpaved trails within areas of suitable habitat. These trails
seldom were used by people, often bisected dense vegetation, and seemed to
be frequently used by all panthers for travel and scent marking. Adult males
frequently crossed highways, and subadult males such as Numbers 10, 24, 28,
29, 30, and 34 also crossed many paved roads during their dispersals. Interestingly, none of these panthers were struck by vehicles.

Female panthers, however., have much smaller home ranges and seem to
avoid paved roads. Of the six females (Numbers 8, 9, 11, 19, 31, and 32) that
had major highways adjacent to their home ranges, none crossed Alligator
Alley and only two (Numbers 11 and 19) occasionally crossed State Road 29.
These roads appeared to form at least one boundary to each female's home
range and may help to explain why females have been hit by cars less frequently than males. In 1993, an aging Female 31 began a series of excursions
away from her consistent home range of at least five years. These unexpected
travels came in the wake of failed reproductive attempts that took her across
State Road 29 and County Road 858 several times before she was hit by a car
and killed on 3 March 1994. Although female mortality due to highway
traffic had also occurred on Alligator Alley, these animals were mostly dispersing young or displaced older panthers. Female 31 and other older females that became highway statistics may have abandoned their home ranges
as a result of social pressure caused by the recruitment of their own female
offspring, which usually do not disperse very far. Experience may increase a
panther's ability to negotiate highways successfully. Of the radio-collared
panthers that were hit by cars, most (Numbers 13, 20, and 31) had little
highway-crossing experience. Others, such as Numbers 11, 12, and 19,
crossed these same roads for a cumulative total of twenty-three years without
incident. (Both Numbers 11 and 19, however, each had at least one kitten run
over on State Road 29.) During a one-year period Number 12 crossed Alli-

gator Alley twenty-eight times, or roughly once every two weeks. Even road improvement activities such as the realignment of State Road 29 and construction of wildlife underpasses along Alligator Alley did not appear to alter panther movements or home range characteristics. On several occasions the tracks of Number 12 were found in underpasses that were under construction, and they could be found winding in and out of parked equipment along a one-mile stretch of State Road 29.

Several researchers have suggested that human activity, increased road density, altered prey density, and removal of stalking cover may reduce the mountain lion's range in parts of the West. All these factors may be operating in south Florida, too, where the combined influence of busy highways and increased human activity may not only discourage highway crossings by resident females but may also eliminate panthers from otherwise suitable habitat. The net result may be a lower female panther population density than the land could otherwise support. The 40,000-acre southern Golden Gate Estates is perhaps Florida's best example of this untapped potential. A crisscrossing network of roads and canals, 20,000 out-of-town lot owners, unrestricted access, minimal enforcement of game laws, and round-the-clock human activities have turned a panther paradise into a zone of avoidance. It was from this highly altered landscape that Panther 9 escaped when a home range vacancy appeared in the Fakahatchee Strand. This was the only case of a resident adult female abandoning an established home range.

The construction of underpasses (Figure 6.3) along Alligator Alley is one of the few solid accomplishments of panther recovery. This highway's conversion to a limited-access interstate with a unique sequence of special bridges for wildlife has significantly improved crossing conditions for a number of species. Tremendous criticism was directed at the state and federal government for spending $10 million in adding these special structures to this two-lane highway's upgrade to an interstate. After all, how would a panther learn where to find these safeways spaced about a mile apart for a 30-mile stretch of new pavement? Telemetry data collected in 1981 and 1982, in conjunction with roadkill locations, were used to pinpoint narrow zones along the highway corridor where panthers crossed regularly. This resulted in the identification of twelve sites between Naples and State Road 29. Another eleven sites east of State Road 29 were added despite the lack of telemetry data, roadkills, or other evidence of panther crossings. Telemetry data indicated that most crossings occurred within the north–south running Fakahatchee Strand, a significant part of the forested system that supports most of the panther population. Apparently the presence of dense forest cover immediately adjacent to the highway has influenced the selection of crossing points by panthers so that crossings have not occurred randomly.

Figure 6.3
Wildlife underpasses help to maintain linkages between forest fragments inhabited by panthers.

The lack of dense forest cover along I-75 east of State Road 29 probably explains why no panthers are known to have used these eleven eastern underpasses.

Clearly underpasses have directly improved conditions for males, which seem to have no aversion to roads. But the females' tendency to use highways as home range boundaries has obscured the benefits to them, at least to this point. Certainly, increased safety to territorial males will increase the likelihood of social contact between the sexes and reduce the probability of local reproductive failure due to diminished encounters between males and females. Moreover, young, inexperienced females will be protected as they disperse from their mothers' ranges in search of home ranges of their own. And this may be critical to maintaining panthers in the Fakahatchee Strand State Preserve—where the recruitment of young panthers likely comes from outside this swamp—and to recolonizing the southern Golden Gate Estates if conditions there can be improved. Given the increasing traffic on Florida highways, underpasses represent the most practical solution to the problem. Improving conditions for panthers where female density could be higher, however, may not be as straightforward. This will require significant changes in the way people use undeveloped landscapes in south Florida.

* * *

Male panthers seemed to be in perpetual motion as they searched for prey, dispersed, or patrolled for mates and potential competitors. During a male panther's first years of independence, his irregular travels could cover over 400 square miles. Number 28 covered more than 130 linear miles between waypoints ranging from Fort Myers to Golden Gate and the Big Cypress Seminole Indian Reservation. Although resident adult males traveled widely, they moved much more regularly throughout their home ranges. Adult females utilized their smaller home ranges with shorter movements and were more likely to localize their activities in relatively small areas, especially when raising young kittens. Average straight-line distances between consecutive daily locations of adult males were more than twice those of females: 3.4 miles for adult males versus 1.4 miles for adult females. The minimum overnight distance traveled by both sexes was zero. Maximum distances were 15 miles for adult females and 24 miles for adult males.

But measuring distances between daily locations just gave us a simplified index of panther movements. Because these movements occur mostly at night, routine telemetry flights did not reveal details of nocturnal activity. To fill this information void, we monitored several panthers from sunset to sunrise in order to describe their hidden nocturnal travels. Shifts in location and activity rates were quite variable among panthers and even within individuals. Their progress ranged from zero to 1.4 miles per hour and occasionally covered circuitous pathways. In reality, panthers move twice as far each night on average than they do between consecutive daytime locations.

Activity level, based on transmitter pulse-rate changes, was generally proportional to total distance moved. But in almost every nocturnal monitoring effort, we observed peaks in activity that occurred without a corresponding rise in distance traveled. Because the transmitters' tip switches were sensitive to any type of up-and-down head motion, it was nearly impossible to associate a particular kind of behavior with variations in transmitter pulse rate. Moreover, a panther could travel long distances without a great increase in pulse rate changes if there was little head motion. Female 19 moved about one-half mile around midnight on 22 April 1991, for example, but her activity level at this time was nearly at its lowest for the night (Figure 6.4). On another occasion Walt and Jayde were monitoring Female 32 on 6 September 1990 when a spike in activity occurred at 10 P.M. without a change in her location. Just after sunrise she walked to a different daytime rest site, which gave them a chance to search her nighttime area, where they found the remains of a freshly killed and eaten armadillo. Tracks around the kill indicated the panther had chased her prey under a stump before digging the armadillo out (probably the activity spike). Similar peaks in activity of other

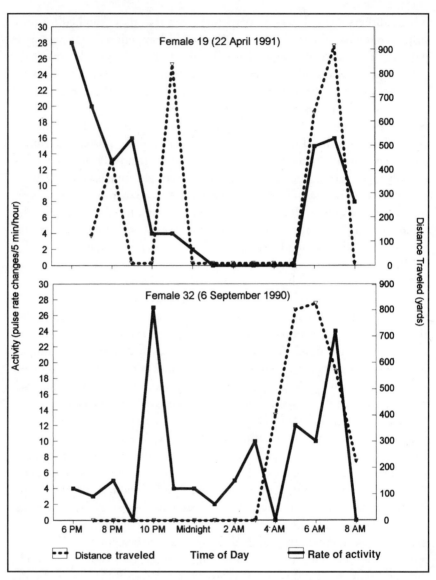

Figure 6.4
Panthers such as Females 19 and 32 could be extremely active without changing location.
Nighttime activity without a shift in location likely reflected feeding. Morning shifts were to
daytime rest sites.

panthers undoubtedly signified killing and eating prey, grooming, and other curt movements that did not result in measurable changes of location.

In general, panthers exhibit a particular twenty-four-hour pattern of activity: leaving the daytime rest site around sunset; stalking or waiting for prey in a small area after dark; killing and feeding on a successful kill; traveling to a new daytime rest site around sunrise (or remaining with a kill); and resting in dense cover during daylight. A panther may not hunt and feed if it travels a long distance overnight. Compared to the twenty-four-hour activity profiles of other large mammals, panthers appear to spend similar amounts of time resting but more time traveling or hunting. This may be a function of the scattered distribution of food in the forested southwest Florida landscape. In areas such as East Africa, prey resources may be clumped and more numerous.

*   *   *

The most challenging aspect of radio-tracking was monitoring the dispersal of young male panthers. During the ten to eighteen months that kittens remain with their mothers, the family unit moves together through the adult female's home range. After males become independent, however, their whereabouts become totally unpredictable as they expand their movements to include more and more unfamiliar territory.

The first panther kitten that we collared opened our eyes to the seemingly random lifestyles of dispersing males. After leaving the Fakahatchee Strand and the care of his mother in November 1986, eighteen-month-old Panther 10 crossed Alligator Alley for the first and only time, squeezed under a chain-link gate, and found himself in the newly constructed Ford Vehicle Evaluation Center, a high-tech facility enclosed by over 5 miles of "panther-proof" fence. His dispersal was one of the shortest on record, however. Number 10, as we shall see later, became our first example of a subadult panther killed by a resident adult. Number 30's dispersal travels included private woodlands south of Alligator Alley just east of County Road 951, mixed swamp forest just north of U.S. Highway 41 near the resort development Port-of-the-Islands, and even included areas of low human density in the northern Golden Gate Estates. In addition, his movements took him away from an area of high prey density to the Fakahatchee Strand. This panther was of interest because as a fourteen-month-old, independent subadult, he was found in the company of an unrelated adult female, Number 31, who was raising her own kittens on the Florida Panther National Wildlife Refuge. This was not the last time we would observe such unexpected interactions, but the unrelated family groups appeared to tolerate the young interlopers. Perhaps young panthers recently separated from the highly social group of the family gained a sense of security. Or maybe they were simply homing in on kills that the unrelated females had made. Number 34, another subadult, continued to travel with his sister for several weeks after their independence, suggesting that family ties are maintained, at least briefly, after kittens leave their mothers.

The actual events that lead up to the separation of panther families can only be guessed. On several occasions we located resident adult males near a panther family just before the breakup occurred. We thought the male's presence had some influence on dissolving the family bonds, but it was unclear if these associations were coincidental or if the resident males forced the kittens to leave their mothers. Black bear families in Florida and elsewhere separate for brief periods during the summer breeding season, suggesting that cubs are not tolerated during courtship. On the other hand, some panther families separated without any apparent influence at all from a resident male. I suspect that the physical condition of the adult female has more to do with this phenomenon than anything else. The presence of males near panther families is likely a function of the adult female's renewed sexual receptiveness. Changes in hormone production most probably caused changes in the female's behavior that signaled her offspring to strike off on their own. This would explain the dissolution of families in the absence of adult males. In any event, without observations that are unobstructed by dense vegetation, this aspect of panther behavior will remain a mystery.

Young males that encounter a female in estrus are also likely to encounter the dominant resident male of the area. After leaving the Ford Vehicle Evaluation Center, Number 10 was killed by resident male 12. Abundant sign of an adult female was found nearby. Number 30 met his fate in the southern Fakahatchee Strand after encountering Female 9 as she courted an uncollared adult male. (Number 37 was captured and radio-collared the next day.) Interestingly, Male 37 sustained no injuries as the result of his thrashing of Number 30, so it is possible that young Number 30 was a submissive victim in this encounter. Over the next few years this turned out to be a commonplace end for young males. In all, six of eleven males that were collared as kittens between 1986 and 1993 were killed by adult males. Only two of these eleven (Numbers 45 and 54) are still alive at the time of this writing.

The longest dispersal from a natal range (measured as the straight-line distance from a known den site) covered over 60 miles and was made by Male 43 from July 1990 to April 1991 (Figure 6.5). After becoming independent at about one year of age, Male 43 traveled east from the Florida Panther National Wildlife Refuge with long movements punctuated by periods of intense local activity. By confining his activity to such a small area, Male 43 may have minimized the chances of interacting with an older male whose presence was advertised with scats, scrapes, and other scent markers. When close encounters did occur, they appeared to urge dispersers along their way—if they survived. Number 43 traveled through the home ranges of resident Males 12 and 26 before concentrating his movements outside of known occupied panther range in the mostly treeless Water Conservation Area 3A, only 18 miles west of Fort Lauderdale. During the fall of 1991, Male 43 retraced his movements westward into permanently occupied forested panther habitat. On 1 November 1991, at just over two years of age, Male 43 was killed by resident Male 26 on the Big Cypress Seminole Indian Reservation.

*   *   *

At the very least, young female panthers in southwest Florida did not exhibit the wide-ranging, unpredictable movements displayed by males of the same age. Nor did they suffer the problems typical of dispersing males. All the young females we captured overlapped extensively with an older resident adult female—likely her mother. Extensive overlap among female panthers was typical in high-quality habitat, a phenomenon that has been commonly observed in cougars elsewhere. Many studies of large carnivores including black bears, tigers, and cougars suggest that females normally establish a home range within or adjacent to their mother's. Female 19 took this generality to an extreme by raising her first litter within her mother's home range before dispersing 10 miles west into the Florida Panther National Wildlife Refuge. Females 48 and 52 left their mother (Number 31) and the same natal range in early 1993. Their respective dispersal distances were 9 and 12 miles. Both gave birth to their first litter by the fall of 1993 when they were each less than two years old and still in their first year of independence. Other young females, such as Numbers 32, 40, and 41, were captured before giving birth to their first litters. Their capture locations, near suspected mothers, suggest that short dispersal distances are the rule for female panthers in southwest Florida.

Dispersal is an important aspect of wildlife ecology. It is thought to have a powerful influence on the stability of populations and their ability to increase. Dispersal of juveniles from their natal ranges is important, too, in reducing the likelihood of inbreeding. Yet even in populations of wildlife species with ample range to accommodate dispersal and an innate compulsion in young animals to disperse, inbreeding can still occur. Despite notions to the contrary, most wildlife species do not appear to have an innate aversion to mating with relatives. Researcher David Mech found that wolves in Minnesota were strongly inbred despite ample opportunities for dispersal. Considering the historical distribution of Florida panthers throughout the southeastern United States, there has been ample opportunity for long-distance dispersal and genetic mixing—not only within the subspecies' range in the southeastern coastal plain but with other cougar populations to the north and west. Nonetheless, after ten years of radiotelemetry research in south Florida, no radio-collared panther had dispersed from south Florida to the north side of the Caloosahatchee River or beyond Lake Okeechobee where there are large expanses of potential range.

Probably no other mountain lion population has been so naturally subjected to inbreeding as the panthers of south Florida. For several thousand years, panthers have been confined within a rigid envelope bounded on three sides by the Gulf of Mexico, Florida Bay, and the Atlantic Ocean. To the north, Lake Okeechobee and the Caloosahatchee River appear to represent substantial barriers to travel (Figure 6.5). Even as recently as a hundred years ago, what little natural exchange took place between south Florida panthers

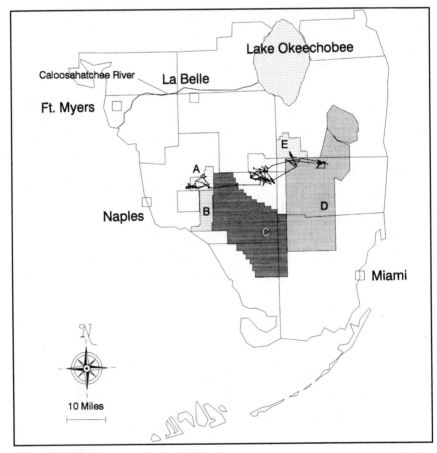

Figure 6.5
Male 43 set a dispersal distance record during his travels to the treeless water conservation areas near the east coast of south Florida. A: Florida Panther National Wildlife Refuge; B: Fakahatchee Strand State Preserve; C: Big Cypress National Preserve; D: South Florida Water Management areas; E: Rotenberger/Holey Land Wildlife Management areas.

and populations farther north most likely occurred through the two land-scape corridors on either side of Lake Okeechobee—the Atlantic Ocean to the east, Lake Okeechobee in the middle, and Lake Flirt (the original head-waters of the Caloosahatchee River) to the west.

Populations with long histories of inbreeding should show little or no in-breeding depression upon further bottlenecking—that is, while features such as cowlicks, crooked tails, and cryptorchidism (only one descended testis) are probably manifestations of inbreeding, Florida panthers seem to have dealt with population reduction and habitat fragmentation surprisingly well. These traits do not appear to have disrupted the social ecology or the demography of the Florida panther, which are remarkably similar to other populations of mountain lions. As far as inbreeding is concerned, mating be-tween closely related individuals has always occurred more frequently in Florida than elsewhere in the species' range. Thus a higher degree of in-breeding resulting in lower genetic variability should be expected in Florida panthers. Perhaps the acceptance of this fact by agencies responsible for man-aging the panther would encourage progress where it is badly needed—in the conservation and management of panther habitat.

* * *

In keeping with their large home ranges, adult males traveled far and wide. A typical resident male, such as Number 12, covered most of his 200-square-mile home range each month. While we have documented these travels through a variety of habitats with radiotelemetry, adult male panthers regu-larly indicated their own presence along game trails, swamp buggy trails, old logging trams, and forest edges via scrapes containing feces or urine. These signals appeared more frequently in the presence of females in estrus and along well-used travel routes. Successful resident males maintained primary breeding rights with females in their home ranges and usually excluded competing males from them. It is not surprising, then, that resident males traveled widely and evenly through their home ranges in order to maintain their social dominance. Obtaining food appeared to be of secondary impor-tance to these males.

Resident adult females were driven by different biological and social needs. With the nearly constant burden of kitten rearing, adult females were usually in search of more prey than was required by a solitary panther. During the months that females spent raising each litter, their movements were initially restricted for two months by immobile kittens. As kittens grew, so too did their ability to travel, allowing the family group to explore more and more of the adult female's home range. Only while a resident fe-male was solitary (that is, seeking a mate or pregnant) did she move widely throughout her home range. During this time the female also made scrapes

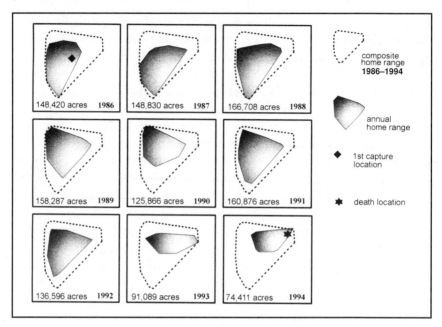

Figure 6.6
Number 12, the archetypal resident adult male, exhibited movement and home range characteristics that were typical for his sex and age class. He maintained a fairly stable home range until old age and competitors overcame him.

to advertise her presence and reproductive status, signals that may have enhanced an encounter with an adult male. Some females experienced as few as five months of solitary travel between litters (the sum of three months gestation plus two months of solitary hunting for denned kittens) before renewing the kitten-training cycle.

Despite the differences in home range use by males and females, the resident adults of both sexes were similar in their consistent use of a well-defined area. Perhaps no other panther exemplified this home range tenacity as well as Male 12 (Figure 6.6). Between his capture in early 1986 and his natural death in 1994, we located Male 12 1406 times within a predictable area that varied little in nine years. The Fakahatchee Strand and the Okaloacoochee Slough formed the core of his home range—a diverse, heavily forested landscape that also supported at least six females who produced about two dozen of his kittens. Although the shape and size of Male 12's home range changed over the years, his movements remained centered north of Alligator Alley on the Florida Panther National Wildlife Refuge.

Though other males appeared and disappeared as 12's neighbors and competitors, he did not respond predictably to each one. From 1986 through

1989, he maintained a large, unchanging home range where he killed Males 10 and 25 as they passed through his territory. In 1990, however, Male 12 abandoned the southern Fakahatchee Strand in response to the appearance of Number 37, the small but aggressive male that killed dispersing Male 30. This withdrawal was only temporary, however, and Male 12 recolonized this part of his former range when Number 37 was killed by a vehicle on State Road 29. But even with its lower prey productivity, the Fakahatchee has always been a magnet for males. Within two years Male 12 withdrew once again in order to accommodate another new male—this time it was Number 51. While Number 51 was establishing himself in the Fakahatchee, the similar-aged Number 46 gradually moved into the northern limits of Male 12's home range. For the next two years Male 12 was increasingly squeezed in a vise created by the two young males, and his home range shrank to 50 percent of the area he used during the 1980s. With Male 12's home range at its smallest, Number 46 killed him in November 1994, ending the longest tenure of any known male panther.

One could infer that Male 12 displayed great wisdom in choosing when to modify his home range in accordance with competitors—during his physical prime, any males that challenged the center of his home range were killed. Males that appeared at the fringes of his home range, however, were afforded minimal stakes of land while Male 12 maintained the breeding rights to most of the females in the area. At thirteen or fourteen years of age, Male 12's physical condition must have reflected the beating it had received over the span of two decades, and at last he gave in to the most recent of many rivals.

Number 46 became the new resident adult in Male 12's former haunts, but his influence was felt further to the east as well. Shortly before Number 46 brought Male 12's long tenure to an end, he managed the same feat with Number 26, the aging resident male of northeastern Collier County. Number 26 had experienced the classic waiting game of a transient male from 1988 to 1990. He spent months at a time in treeless, shrubby, and wet expanses in the western Everglades, making forays across the L-28 drainage canal (apparently using a water control structure near the Big Cypress Seminole Indian Reservation to get across). After the death of resident adult Male 17, he became the reigning male of this productive home range. But he was not spared the constant reminders of other males-in-waiting. From 1991 through 1994, Number 26 was surrounded by younger transients: Males 28, 29, 34, 43, and 45. Four of these cats succumbed to natural deaths by 1993 (two in fact were killed by Number 26), and only Male 45 (an offspring of Panthers 12 and 19) survived through 1994. Then Number 46 showed up in 1993 and killed Number 26 in July 1994 before doing the same to Number 12.

Within the span of four months Number 46 had eliminated two long-standing residents—events we had not witnessed in over a decade. The consequences of this mortality would not be damaging to the panther's resilient social structure; reproduction was guaranteed to continue because surplus males are constantly available to fill the rare voids that occur. Male 46, with his unknown pedigree, will have the opportunity to breed with the females previously available to Male 12; Male 45 will likely inherit the land and females within Male 26's home range; and a new cast of transient males will set up their tentative holding patterns.

Female panthers did not exhibit the home range dynamics typical of males, but they too became attached to one area. Within eighteen months of their birth, females establish home ranges and begin reproducing. Annual changes in female home range shapes and sizes were due to the changing demands of growing kittens. Age is another factor in the female's home range stability. Female 31 consistently used her home range from 1989 through 1992, during which time she raised seven kittens. In 1993 she began a shift to the north that embraced an area of scattered woodlands that was crossed by a busy highway where she was finally killed. Her failed reproduction in 1993 and 1994 may well have been a reflection of declining health due to advancing age. Perhaps her uncharacteristic wanderings were somehow related to this decline. And because at least two of her kittens (and possibly two others we did not collar) had established home ranges adjacent to her, she may have been responding to social pressure created by the success of her own productivity. Just as Males 12 and 26 were immediately replaced by younger males, Female 31 had already seeded the landscape with more than enough productive females to carry on in her stead.

It is clear, then, that resident adult panthers became attached to a well-defined plot of ground and adhered to it tenaciously until circumstances beyond their control intervened. These circumstances were usually related to age and health, but occasionally a younger male like Number 46 would succeed in displacing an older one—thus short-circuiting a process that most males waited many years for, often with no reward for their patience. A panther demonstrates its residency with a consistent pattern of occupation—a pattern that pertains not only to individuals within home ranges but to home ranges within the landscape as well. In other words, a resident panther occupies an address just as people do—and when the panther is evicted or dies, the address remains the same and a new occupant moves in. Sometimes the size of the "yard" changes, but the central "homesite" remains constant. I formed this concept of the panther address from telemetry data collected in the Fakahatchee Strand from 1981 through 1994 (Figure 6.7). During this time, at least five adult males lived here. Even though there was obvious vari-

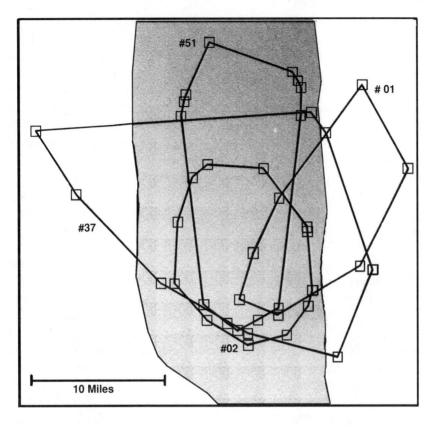

Figure 6.7
When an adult male died, another always showed up to take his place and use essentially the
same home range. This pattern was most evident in the Fakahatchee Strand (*shaded*), where
panther research has gone on since 1981.

ation among these cats, there was also a remarkable symmetry to the centers
of their home ranges. Only Male 1 was an obvious outlier, but he was a decrepit male when captured and was compelled to leave the preferred habitat
for reasons most likely related to old age, failing health, and the influence of
his neighbor, Male 2. The others centered their activities along the backbone
of the strand, one of the most distinctive landscape features of south Florida.
This central tendency within a seldom vacant address has been maintained
for at least two decades. Identical patterns most likely hold true for the rest of
panther range.

* * *

As their name implies, transient males are not permanent residents. Although they may use a large tract and remain attached to it with more fidelity than a subadult male, they seem unable to remain in one area for more than a few months at a time. These were the cats that waited in the wings for a resident adult male to keel over or become decrepit enough to be pushed out of the way. Even transient males replaced each other as they waited in line for the next best opportunity to move up the ladder of social dominance. With the exception of Male 20's literal scrape with death by pickup truck three months after his capture, his pattern of movements was virtually identical to that of Male 13, another transient we had captured the year before. These male panthers were captured at opposite ends of their north–south running home ranges, and although the area used by each was pretty much the same, its occupation was separated in time by one year. In other words, Male 20 took over where Male 13 left off.

Number 13 spent the last year of his life in a small area centered on northwestern Bear Island and a small adjacent private ranch where Female 11 was raising her growing litter. As these kittens grew, Male 13 was increasingly in the company of Female 11 and at least one of her kittens, Number 19. Based on the published literature on mountain lions, we assumed that neither of these females were sexually receptive. Research suggested that females do not breed until three years of age and that adult females do not breed again until their young become independent at about eighteen months. If females were scarce relative to adult males, then maybe Male 13 was simply maximizing his chances of breeding by being on hand when this panther family broke up and Female 11 became available again. But if this were true, Male 13 was also taking a big chance by squeezing himself in between the home ranges of two resident adult males, Numbers 12 and 17. Male 13 successfully avoided these older males, but when he crossed State Road 29 on a late December night in 1987 he was struck and killed by a vehicle.

Within a matter of weeks, Male 20 abandoned his temporary home range to fill the space that his predecessor, Male 13, had just unwillingly vacated. Male 20 had waited at least a year, despite his 20-pound weight advantage, until the more senior albeit nonresident male disappeared. Male 20 would eventually sire Female 11's next litter, but his success would be just as fleeting as Male 13's. He died in his new smaller home range on 24 August 1988 of natural causes that may have been hastened by canines worn down during his brief stay in captivity. Despite Male 20's recovery from road injuries and return to his home range, the handling of injured animals was still a developing science at the time. Even though he was kept sedated, Male 20 constantly attempted escape from his Miami Zoo enclosure by gnawing its concrete footings. Despite an exceedingly caring and competent staff, our wild panthers did not respond well to their new confinement. The result for Male 20 was canines that were eroded to nubbins compared to his original, dagger-sharp equipment. There is little doubt that his killing and feeding abilities were compromised to the point where his nutrition suffered.

* * *

The Florida panther's ability to survive as an integral part of Florida's native fauna depends on successful reproduction. Early on in panther studies, researchers assumed that the Florida panther population did not regularly produce kittens. Perhaps, like the south Florida woodstorks that require precise amounts of rainfall at just the right time of year for successful nesting, good panther reproduction occurred only under exceptional environmental conditions that might happen just once a decade. But close monitoring of adult females helped us dispel the myth that the panther population was composed of old, infrequently reproducing individuals. Our research revealed behavior patterns that were unique to females raising kittens.

Females maintained dens for forty-seven to fifty-six days while their kittens were incapable of independent travel. The end of denning occurred when kittens had grown sufficiently to accompany their mother to kills. We monitored four females for 4624 continuous hours during each of their critical two-month denning periods. Although they were most likely to be found at the den between 8 A.M. and 4 P.M., during these two months they were with their kittens only half the time. Den arrivals centered around 8 A.M. and departures centered around 10 P.M., though these events might occur at any hour and varied from female to female. Female activity at den sites peaked immediately before departures and after arrivals and probably reflected greeting, grooming, and nursing behavior.

When we compared hourly averages between four denning females and five solitary panthers of both sexes we found the same cycle of twenty-four-hour activity revealed by telemetry flights. But the activity of females at dens was lower during evening hours and higher during daylight than that of solitary panthers. I suspect that behavior related to kitten care at the den led to increased activity while solitary panthers were normally motionless and at rest in daytime beds.

* * *

In long-lived animals such as the Florida panther, reproductive patterns have to be studied for years before generalities can be made about the population's potential to increase. Since telemetry work began in 1981, fourteen radio-collared adult female panthers have provided a glimpse into this facet of their biology. For a wildlife population to maintain its numbers over the years, females must produce enough offspring to replace losses of resident adults. This does not mean that every female kitten must survive. It just means that the adult female must replace herself and a male before she dies. Growing populations produce successful dispersers that colonize new territories or are able to coexist with standing residents as the result of improving habitat conditions. A declining population is incapable of offsetting mortality.

In over ten years of study in southwest Florida, ten panthers produced at least forty-eight kittens, versus a loss of twenty-six radio-collared panthers. In other words, Florida panthers have produced nearly twice the kittens necessary to replace lost adults. But because the loss of resident adults is slow, there are few opportunities for young panthers to establish home ranges. Among collared kittens or subadults, only three females (Numbers 19, 48, and 52) and one male (Number 28) were recruited into the population as resident adults between 1986 and 1993. Wide-ranging movements were universal for male panthers following independence from their mothers. Young female panthers, however, were recruited readily into the population without a long dispersal. A long-distance dispersal and an age of at least three years seemed to be prerequisites for male reproduction. Luck, however, is probably the biggest factor influencing the establishment of a male as the dominant resident of an area.

Age at first reproduction is perhaps the most important parameter of demographics in a population. But since there is no reliable technique for determining the age of adult panthers, we often found it impossible to pinpoint the age at which their first successful reproduction occurred. With black bears it is a simple matter to extract a small, unimportant tooth and then assay its growth rings with the aid of a microscope. Deciphering panther ages, however, is possible only by capturing dependent kittens of known birthdates and monitoring them via radiotelemetry. The age we assign to an adult is based on its physical appearance when it is captured for the first time—at best a subjective estimate. Although we have yet to observe the successful recruitment and breeding of a male panther first captured as a kitten, all three females captured as kittens reproduced before they were two years old. Females 19, 48, and 52 conceived their first litters at eighteen, sixteen, and twenty-one months of age, respectively. The average of these first reproductive events (18.3 months) is nearly 1.5 years earlier than predicted by studies of western cougars. These record-setting females resided in areas of abundant prey, and their home ranges overlapped with their mothers'. Similar patterns of overlap have been seen in female tigers in Nepal and female black bears in Minnesota. Our observations of the female maturation process suggest that Florida panthers possess a high reproductive potential. Their population growth is limited only by habitat quality and availability.

Even though panthers are physiologically capable of reproducing throughout the year, more litters were documented between March and July than in the remaining seven months combined. Female panthers are induced ovulators—that is, they require courtship and copulation before mature eggs are ovulated for fertilization. While mating, wild adult male and female panthers spend from two to ten days together. Our preliminary data suggested that a distinct breeding season occurred during winter, but as we began to observe additional litters, this peak widened to include most of the calendar year. Nonetheless, a higher rate of sexual encounters during late fall and early winter results in a pulse of denning from early spring to summer. This pattern may be a response to environmental fluctuations ranging from daylength to food availability. Inasmuch as water levels are lower, temperatures are cooler, and most deer fawns are produced in late winter and early spring, panther kittens produced at this time would seem to have a higher probability of survival. A secondary peak of denning appeared during the fall—a phenomenon we would have missed if fieldwork had ceased after five or six years. This lesser pulse of reproduction may reflect rebreeding of females that lost litters earlier in the year.

Despite the huge landscapes in which panthers live, social contact among these solitary cats apparently allows residents to know the sexual identity and condition of their neighbors. The high incidence of scraping by estrous females suggests that the chemical substances known as pheromones play a role in successful mating. Pheromones constitute an important form of social communication in many mammalian carnivores. Skunks have taken this glandular function to the extreme by turning it into a potent defense mechanism. In panthers, estrogenic hormones or related chemical cues that are deposited along with a receptive female's urine and feces may be a powerful stimulant to mature males. Thus the female's reproductive condition may affect not only male behavior but male reproductive physiology as well. Female panther pheromones are likely responsible for rises in testosterone, increased sperm production, and aggressive behavior in adult males. The often lengthy encounters between copulating panthers, up to ten days, may help to ensure that sperm are of sufficient quality and quantity and inseminated before the female's period of fertility passes. Gestation in panthers averages 94.6 days. The average litter size is two and can range from one to four. These figures are conservative, however, because we observed most kittens after they were four months of age. Surely some neonatal mortality occurs, though we did not find evidence of this. Between 1992 and 1994, the six litters of neonates we handled within two weeks of their birth averaged two kittens with an even sex ratio. Because the initial litter size is similar to the size of six 12-month-old litters we studied, it appears that little mortality takes place between one and 52 weeks of age.

Like most other felids, Florida panthers have a well-developed ability to recycle quickly after a litter is lost, or following independence of kittens. But unlike other wildlife species such as North American bears and migratory birds, panther reproduction can take place year-round. By avoiding the restrictions of a distinct breeding season and living fairly long lives, panthers have demonstrated that they are capable of withstanding occasional reproductive failure. Combine this with the good reproductive rates they exhibit, and we are forced to conclude that today's Florida panther is able to increase its numbers rapidly.

* * *

Given the high degree of home range overlap among radio-collared panthers (up to 100 percent for females), the opportunity for social interaction is great. Even so, our data rarely located two or more panthers in the same spot simultaneously. Interactions between two adult females, two adult males, one adult male with one juvenile male, and one adult female with one subadult male were especially unusual. These combinations accounted for less than 1 percent of all documented interactions. Almost all social interactions were between adult females and their kittens. But even these pairings occurred only about 20 percent of the time that we located individual animals from the air during morning flights. Thus the basic social unit was frequently dispersed. Interactions between adult males and females were second in frequency, but these events were even less common—except for Female 18 and her interactions with several adult males (among them Numbers 17, 20, and 28). Out of sixty simultaneous locations of adult males and adult females during 1986 and 1987, interactions between Female 18 and her suitors accounted for nearly 50 percent of them. Female 18 did not reproduce in three years of monitoring, however. Recurring estrus without pregnancy may have explained her extensive home range, her wide-ranging movements, and her frequent interactions with males.

Despite the rarity of some forms of interaction, all were important. Interactions between males, for example, usually resulted in serious injury or death. Most independent male panthers captured in southwest Florida exhibited numerous scars and other injuries in various stages of healing. No aggressive encounters were documented between adult females, however, even though they were occasionally found close together. In the absence of a high rate of unnatural mortality, aggression between males appears to be by far the most common form of panther death and a key determinant of male spatial and recruitment patterns. Overall, panthers in southwest Florida exhibited an annual mortality rate of about 17 percent, which is comparable to estimates of mortality in unhunted mountain lion populations in the western

United States. Further, while panther kittens have been portrayed as the victims of high mortality, our examinations of young kittens in their dens and our captures of older kittens for collaring suggest a survival rate of between 80 and 90 percent through their first year. Ample reproduction, early maturation by females, longevity of resident adults—all suggest that panthers are more than capable of replacing themselves. In fact, given the surplus of kitten production, the Florida panther has the potential for population increase—if and when new habitat is made available.

PANTHERS
IN THE LANDSCAPE

Early fieldwork from 1981 through 1985 concluded that the Florida panther is a creature of deep and remote swamplands such as the Fakahatchee Strand. While such habitat is indeed an important feature in the south Florida landscape and panthers are often associated with these forests, our study began to suggest that panthers are not in fact creatures of wetlands. In any case, we would have preferred these mosquito-ridden and poisonous-snake-infested swamps over our climate-controlled office. The hours we spent outside in remote and beautiful settings were invaluable. Not only did we get to know individual study animals, but these outings were often the only incentive we had for putting up with the many distractions of the office. Even so, it was the time we spent in the sterile confines of air-conditioned cubicles that allowed us to realize the value of our sweat and toil. In between phone calls from the media, interruptions from Tallahassee administrators, and innumerable meetings, we were gradually able to make sense of telemetry data and begin to understand the panther's place in the south Florida landscape.

The aspects of natural history that most clearly reflect the panther's use of the landscape have been described in analyses of home range and habitat use. Home range was defined by W. H. Burt in 1943 as "that area traversed by the individual in its normal activities of food gathering, mating and caring for young." Habitat is an aspect of home range that includes the environmental variables, such as vegetation and topography, within which panthers interact, feed, rest, and reproduce. Studies of wildlife before the development of miniature, durable electronics depended on repeated captures of marked individuals over an extended period and large area in order to outline the boundaries of home ranges. Other descriptions of habitat use have relied on the discovery of tracks and other sign; still others depended on direct observations of the study animal. With the advent of radiotelemetry equipment in the early 1960s, it became possible to monitor study animals in almost any location, at any time, from observation distances that did not alter their behavior. Not only can repeated trapping or direct observations influence animal health and behavior, but these methods do not offer the logistical advantages provided by telemetry. (In fact, telemetry locations are still often referred to as "recaptures" even though the animal is not touched or disturbed in any way.) This technology allowed us to collect the most accurate information possible concerning panther habitat use without constant intrusion into their lives.

The home range of each panther was defined by a minimum convex polygon (as in Figure 7.1). We drew the boundary of this area by connecting the outermost points on a map depicting each panther's telemetry locations. After plotting the areas used by panthers, we could examine discrete aspects of the environment that seemed important to them. By identifying the specific habitat type where panthers were located during each flight or ground inspection, we got a good picture of how they used the space available to them. Perhaps more important, we could also see which habitats they preferred.

* * *

Adult panthers had annual home ranges that varied from 20 to 456 square miles. Home ranges averaged 200 square miles for resident adult males, 75 square miles for adult females, 240 square miles for transient males, and 67 square miles for subadult females. To put these figures in perspective, consider the contrast between the distributions of people and panthers. New York City in 1990 covered 309 square miles and had a human population of 7,322,564. Not only do these millions of people live within an area equivalent to one slightly larger-than-average adult male panther home range, but each city resident can personally claim less than one-third of an acre of the metropolis. By contrast, about seventy-four panthers inhabit 1945 square miles and each can claim more than 26 square miles. Certainly most New Yorkers are not restricted to their tiny portion of the city—they travel freely about on foot, car, and mass transit—but relative to panthers they are apparently much more tolerant of each other (crime statistics notwithstanding). That is, human home ranges in Manhattan exhibit tremendous overlap whereas panthers live as virtual landscape hermits.

Adult female panthers appeared tolerant of each other and, relative to male panthers, exhibited extensive overlap. Resident adult males tended to avoid each other (Figure 7.1). The largest adult female home range was that of the nonreproductive Female 18. Female 8, another nonreproductive panther living in the Fakahatchee Strand, had the next largest female home range. Perhaps the lack of metabolic burdens due to pregnancy and kitten

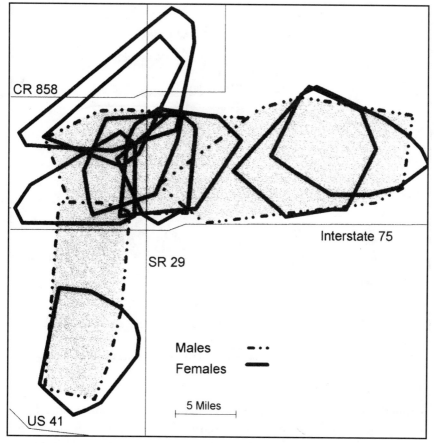

Figure 7.1
Male home ranges (*shaded*) were two to three times larger than female ranges and overlapped little with other males. Females seemed more tolerant than males of panthers belonging to their own gender.

rearing permitted these females to move widely—in some cases, more widely than males. And without kittens acting as home range anchors, these females may have had an easier time finding prey. The largest home range we encountered was that of Number 24, a young transient male monitored for seven months in Glades and Highlands counties (Figure 7.2). Although he was the only panther to have been radio-tracked on the north side of the Caloosahatchee River, his presence there meant either that other panthers resided there or that dispersing males occasionally crossed this large watercourse. At the very least, Number 24 showed there was panther habitat outside of southwest Florida.

While it is impossible to predict the exact location of resident panthers

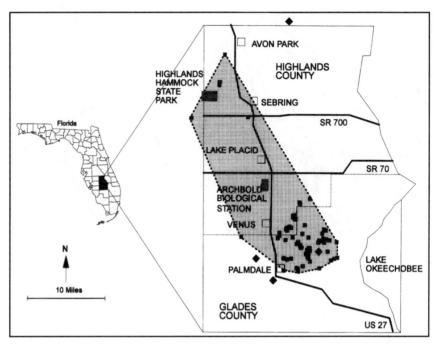

Figure 7.2
Male 24 was the only Florida panther monitored north of the Caloosahatchee River. He died
mysteriously seven months after his capture.

within their home ranges (except when females and kittens are at natal dens),
they use these areas with enough regularity that they can always be found.
Even if a radio collar had ceased to emit a signal, the animal was eventually
recaptured because we knew its habits of movement so well. Females exhibited the greatest stability in home range use. Only death seemed capable of
wresting a resident female from her home. Female 9, however, made a large
shift after she raised her second litter in the northern Fakahatchee Strand
State Preserve. Prior to this, from at least March 1984 through April 1987,
resident Female 8 resided in the remote southern Fakahatchee Strand before
she was taken into captivity due to poor health and lack of reproduction (as
the result of old age). Within one year of Female 8's removal, Female 9 had
abandoned her home range in the southern Golden Gate Estates and reestab-
lished herself in the older female's now vacant home range. The geometric
center of Female 9's new home range was about 9 miles south of her 1986 and
1987 home range, and it closely mirrored Female 8's old home range. Despite
Female 8's lack of reproduction and poor physical condition, she maintained
the exclusive rights to the area—apparently the best female panther habitat
the Fakahatchee Strand had to offer. The superior quality of the southern

home range was suggested by Female 9's subsequent rearing of two healthy kittens in 1991. Her previous two litters had consisted of only one kitten each. Sometimes adult females would shift their established home ranges for a while. Births signaled a drastic reduction in a female's home range use as a result of the care required by helpless kittens at the den. Denning lasted for up to two months at which time the kittens began traveling with their mothers. With growing, hungry kittens, each female shifted to different parts of her home range, apparently to avoid undue pressure on the local deer population.

Male panthers maintained larger home ranges than females, and their occasional shifts were considerably more extensive. In every case, the shift of a resident male could be attributed to the presence of a neighboring male. Number 7, captured as an adult in the eastern Big Cypress National Preserve in 1982, used a 253-square-mile home range for over one year. After his radio failed in May 1983, he was recaptured in January 1985 about 25 miles to the west in the Fakahatchee Strand. Two months before Number 7's reappearance and recapture, Number 2, an old male (more than ten years old), was found dead in the Fakahatchee Strand, apparently killed and fed upon by another panther. Another old male, Number 4, shifted his home range from the northern Fakahatchee to more than 6 miles north of I-75 during the interval that Number 7's transmitter was not functioning. I suspect that the younger, healthier male forced Number 4 to shift his home range to avoid a confrontation, but that Number 7 subsequently killed Male 2 when he failed to leave. Male home range shifts, whether large or small, occurred frequently, underlining the dynamic spatial arrangements of this small population.

Permanent home range shifts in adult panthers were usually related to the deaths of residents. Older animals were more susceptible to home range usurpation than panthers in prime condition, but even healthy residents were susceptible to injuries or death resulting from accidents. Nonetheless, between December 1985 and March 1994, no reproducing females were forced from their home ranges and only three resident males are known to have died. In every case, they were quickly replaced. The fabric of home range dynamics known as the land tenure system seems to have evolved to encourage the occasional shuffling of individuals while maintaining an overall pattern of order. The land tenure system in panthers is characterized by extensive home range overlap among females and minimal home range overlap among resident adult males. Despite the pressures placed on panthers by urban development and agricultural growth, they maintained a surprisingly stable distribution of home ranges in southwest Florida. In fact, their home ranges fall well within the size limits exhibited by this widely distributed species. As a rule, home range stability is typical of unhunted felid populations operating in a land tenure system. Such stability is probably important in maintaining uninterrupted reproduction in resident females, dividing available food resources, and reducing strife among males.

* * *

Native upland forests, especially hardwood hammocks, were sought out by panthers to the exclusion of all other vegetative communities. Pine flatwoods, cypress swamps, and cabbage palm woodlands were used by panthers, too, but not nearly to the degree they used hammocks. Agricultural lands, freshwater marsh, thicket swamp, and mixed swamp were used only in proportion to their availability or else they were avoided. These habitats are apparently perceived by panthers as too open, too wet, or both. Number 24, the transient male captured and monitored briefly in Highlands County, used most habitats in similar proportions to his relatives living south of the Caloosahatchee River. But because of the very different geological history of the area, he was the only panther known to use bay forests and sand pine scrub, two plant communities that are absent or rare in south Florida.

Soil and vegetation maps of south Florida illustrate clearly the habitat changes from the southern Lake Wales Ridge, where Number 24 roamed, south to Florida Bay and the Everglades. Generally soil fertility, forest cover, and habitat diversity decrease while treeless, inundated habitats increase from the northwest to southeast. That panther abundance exhibits a similar pattern is no coincidence. And it is no coincidence that panther abundance declines sharply south of I-75. The most frequented parts of panther range are associated with large, distinctive forest systems forming the panther's population core. In south Florida, these forest systems include the Fakahatchee Strand, Kissimmee Billy Strand, Corkscrew Swamp/Flint Pen Strand, Gum Swamp, Okaloacoochee Slough, and Picayune Strand. None of these forest systems crosses into the relative distribution vacuum in southern Big Cypress National Preserve or the Everglades—suggesting that large extensive forests are important in explaining panther distribution. Interstate 75 just happens to traverse the southern edge of this dense forest where clearing and road construction were more easily engineered.

Even in individual panther home ranges, peaks in the frequency of radio locations demonstrate that preferred habitat is not only sought out but is unevenly distributed. Forest systems south of I-75 and east of State Road 29 are much smaller, narrower, and more fragmented than the swamps and upland forests of northern Collier County and southern Hendry County. Just as many of Collier County's cypress domes have been orphaned by the isolating effects of sprawling agriculture (Figure 7.3), the small patches of forest typical of most of the Big Cypress National Preserve are naturally surrounded by nonforested habitats that panthers usually avoid. Thus habitat conditions in eastern Collier County and the Everglades are not capable of supporting a permanent reproducing population. An examination of the few home ranges that have existed in this part of south Florida suggests that panthers must work much harder to make a living. Not only are forests more scattered, so too are potential prey. Home ranges of the two resident adult females living in the eastern Big Cypress National Preserve were two to four times larger than the average female home range in better habitat to the north. Fragmented forest cover, lack of preferred habitat, an overabundance of wetlands, and sparse prey populations have apparently made the preserve mostly poor panther habitat. In terms of its value to the population, this area acts as a population sink. None of the panthers captured south of Alligator Alley and east of State Road 29 have successfully dispersed northward to reproduce. And even panthers that survive the trip would find the better-quality habitat already occupied by residents.

Hardwood hammocks have been preferred by all panthers studied since 1981. This habitat type generally covers no more than 50 acres in any one location and is usually contained within larger tracts of cypress, mixed swamp, and pine flatwoods. Pine flatwoods were used frequently by panthers and became all the more important because most females chose to give birth in this

Figure 7.3
Fragmentation—whether natural or, in this case, unnatural—reduces the likelihood that a panther population will use a particular patch of forest.

still widespread habitat. Thicket swamps, freshwater marsh, and agricultural lands were probably avoided due to inadequate cover or seasonal flooding. Although some western mountain lion populations exhibit extensive habitat shifts due to seasonal movements in prey such as mule deer and elk, habitat use by Florida panthers was similar for both wet and dry seasons. In southwest Florida, deer do not undertake seasonal home range shifts and panthers do not respond measurably to fluctuations in water levels.

*   *   *

Although our regular flights to locate panthers gave us a basic understanding of habitat use patterns, it took more intensive work on the ground to determine exactly how they were using their favorite haunts. Using a radio signal as a homing beacon, we made countless forays into the woods in order to inspect the vegetation chosen by panthers after a night of hunting. In nearly every instance we approached to within a few yards but never caught a glimpse of the tawny animal as it quietly slipped away. Infrequent sightings were memorable experiences, but this was not the purpose of our intrusions. Instead, we noted the plant species and structure of the beds where panthers spent a large part of each day. Gradually a picture of preferred daytime cover emerged—an image that helped to explain the general patterns of habitat use that were revealed by our flights.

We learned that panthers seek thick, concealing cover that provides both a vertical and a horizontal screen. Fern beds, cabbage palms, fluted tree trunks, and limestone solution holes were occasionally the dominant features providing daytime cover and even maternal dens. But the most commonly used plant species was the ubiquitous saw palmetto. This low-growing palm is usually associated with pine flatwoods and hardwood hammocks, where it thrives on well-drained, sandy soils. It can grow as scattered individuals among live oaks or form extensive thickets in association with slash pine. It was in these thickets that we found panthers two-thirds of the time. Usually the only evidence of panther presence was an oval of bare earth bordered by the creeping rhizomes of the palmettos. On close examination, a few shed hairs and a crawling tick or two (dislodged by grooming) confirmed that a panther had just left.

The saw-toothed stems of palmetto leaves formed an interlocking canopy that kept temperatures cooler in summer and warmer in winter, but they also shredded human skin and clothing as we tripped and crashed into these tangled refuges. Our difficulty in groping our way through these jungles may not be unique. Other wildlife species, including potential predators of helpless kittens such as black bears, probably have trouble making their way through palmettos without alerting resting panthers to their presence. My first plunges into palmetto thickets were accented with anxiety—not so much from a potential encounter with a ferocious animal, but from the scorpions, rattlesnakes, and other biting creatures that Florida folklore claimed to infest them. As it happened, the most commonly encountered denizen turned out to be black vultures, which, like panthers, preferred palmettos for nesting. In these cases, the only aversion was the foul odor of the vulture nest. Poisonous snakes and stinging arachnids were never a problem.

Our studies revealed another advantage conferred by palmettos. Although we always ended up finding panther dens, our searches were hampered by the kittens' fur color, which matched the immediate environment. Palmettos used by panthers sprout a verdant, rooflike canopy that is usually 6 to 8 feet tall. At ground level, however, dry sand, palmetto stems, old palmetto flower stalks, and dried palm fronds create a subdued world of browns, grays, and deep black shadows—the perfect background for concealing the drab, black-spotted pelage of two-week-old panther kittens.

In light of the amount of time that panthers spend in palmettos and the reasons they stay there, it is no understatement to say that the saw palmetto is the single most important plant species for the Florida panther. Saw palmettos are also used regularly as food and cover by white-tailed deer, wild hogs, raccoons, and wild turkeys. Black bears feast on palmetto fruit, and they, too, make winter dens among their roots. Literally hundreds of vertebrates and probably thousands of insect species depend on saw palmetto during some part of their lives. Although this plant is often the bane of land managers because it is extremely flammable and takes up precious space for cattle, saw palmetto deserves special attention as an important habitat for much of Florida's wildlife. For the panthers, however, saw palmetto is of such importance that the suitability of their landscape is directly proportional to the distribution and abundance of this plant.

It is tempting to think of panther habitat only as the trees, shrubs, grasses, and sedges that carpet the surface of south Florida. But the distribution and abundance of panthers may have as much to do with other wildlife and soils as the plant life. Soil texture, moisture, and elevation are directly responsible for the patterns of plant distribution that have molded our concept of ecological communities. These same factors influence the productivity, lushness, and palatability of plants and how they are used by a variety of herbivores ranging from caterpillars and fruit flies to swamp rabbits and white-tailed deer. Just as a black bear may spend a week in a 2-acre woodlot gorging and fattening on the fruits of saw palmetto, a panther can afford to lead a life of relative ease where prey is abundant. Working less for more allows some panthers to have smaller home ranges and reproduce at higher rates than panthers living where food is less abundant. In the sense that more food and less work equals larger body size and more progeny, a panther is what it eats.

*   *   *

As hunters, panthers differ from the gregarious African lion and fleet-of-foot cheetah. But they are similar to most other wild cat species because they hunt alone under the cover of darkness and in dense forest. For Florida panthers, dense vegetation is important not only for day beds and denning but also for stalking cover, where a close approach to a deer, hog, or other potential meal precedes a short, quick, and usually lethal lunge. In this respect, the Florida panther's hunting methods are similar to those of leopards, tigers, jaguars, bobcats, and even house cats. Panthers usually kill large prey such as deer and hogs with a bite to the neck, head, or throat. Then the carcass is dragged to the nearest dense cover where feeding begins. In most of the cases where we found a fresh kill, the entrails were removed and buried nearby and the carcass was covered with the nearest vegetation. Panthers appear to feed on the vitamin-rich organs such as the heart, lungs, and liver before consuming muscle tissue. Covering their kills can sometimes keep the meat fresh enough for two or three days of feeding, but the high temperatures in south Florida usually render them inedible long before this. In cold western climes, mountain lions will feed on their kills for as long as two weeks—and other species may feed on the frozen carcasses all winter. The implication of rapid food spoilage in Florida is that panthers must make kills more frequently than their relatives in western states.

Because of the dense nature of southwest Florida's forestlands we never watched a panther make a kill. During the fall of 1987, however, the panther's killing behavior was made audibly clear to Jayde and me as we approached Number 9 on foot along an old logging tram through the dense mixed-swamp vegetation of the Fakahatchee Strand. Number 9 had recently given birth to her second litter, and we were anxious to discover how many kittens she had with her. While searching for tracks along the trail we startled a yearling white-tailed deer as she pulled swamp lilies from 10 inches of water. Suddenly the panther's radio signal thumped from the receiver. She was very close. The deer reacted to our presence by quickly lifting her dripping head from the tannin-stained waters and then bounding away in the direction of the panther. She disappeared into a willow thicket but the syncopated splashing of her hooves could still be heard. Without warning, the deer's retreat came to an abrupt halt as a loud crash erupted from the thicket. A brief moment of silence preceded three bleats of deer distress, then silence again.

The entire event lasted about fifteen seconds. Not wishing to risk spooking Number 9 away from her suspected kill, we left the scene and returned the next day to seek the physical evidence of what we had heard. At the base of a large laurel oak were the remains of the deer. Nothing but long bones and a few scraps of hide remained. Number 9 and her female kitten (now verified by dainty tracks at the kill) had consumed every scrap of edible flesh on the small deer. Whether or not Jayde and I had influenced her hunting success will remain a mystery, but Number 9 and her kitten took full advantage of the opportunity to kill a fairly large prey item.

Small prey such as raccoons and nine-banded armadillos likely are taken opportunistically by traveling panthers. Sometimes field sign allowed us to trace a panther's detour from a trail to dig out and kill an armadillo, or another's ambush of a napping deer. But these discoveries occurred too infrequently for us to describe the panther's diet with confidence. The only practical way of collecting information on food habits was to identify prey remains in the feces (scats) we were able to find along trails used by panthers. Our efforts to describe food habits were of great importance because of two widely held notions—that Florida panthers were underfed and starving and that panther habitat was a constant and did not change with geography.

We collected the scats of Florida panthers during routine field activities or during annual capture attempts. With enough practice, an alert observer on a moving swamp buggy could detect a scat from 20 yards away and identify the contents from half that distance. Besides being quite large, panther feces are particularly odoriferous, so they tended to accumulate in frozen stacks of ziplock bags in our laboratory freezer. Apart from their repulsive scent, fresh scats can also be particularly hazardous to handle. If carried too close to the skin, as some of us found out, a bag containing the remains of a panther meal can erupt with hatching hookworms that respond to body heat before burrowing through the plastic and entering the nearest human epidermis. Although people are not a natural host of this parasite, hookworms can leave burning and itching trails under the skin. Treatment is slow and uncomfortable.

From time to time we would spend a day closely examining the hair and bone fragments contained in our most recently thawed scat collection, using field guides, hair collections, and reference bones to help us reconstruct the diets of our study animals. Kills were found much less frequently, but for our purposes this was not much of a loss. An impression of panther food habits based on kills alone would have overemphasized larger prey. This is because smaller prey items such as armadillos and rabbits are consumed nearly in their entirety—what little is left of them is next to impossible to find. Because scats are a direct reflection of diet, even the smallest of meals, they provide the least biased sample for understanding what panthers eat.

We found only thirty-eight kills between January 1986 and September 1989, but 270 scats had been collected between January 1977, when field searches first began, and September 1989, when we concluded our food habits analyses. Scats were found primarily along wooded trails that panthers regularly used. Males were particularly prone to leaving scats on mounds of pine straw as evidence of their passing. Seven vertebrate species were identified as kills; fourteen species were identified in scats. No seasonal changes were noted in the panther's diet, but individuals living north of Interstate 75 consumed more hogs whereas those living south of this highway consumed more raccoons. This geographical gradient mirrored the distribution of heavy forest systems—a pattern reflected in the soil as well. Not only does the interstate follow the edge of the panther's preferred forest. It also falls on the natural separation between better-drained, higher-fertility soils to the north and frequently inundated, less productive soils to the south. Soil scientist H. Yamataki described the predominant soils of the Fakahatchee Strand—the Boca, Riviera, and Copeland fine sands—as being under water for six to nine months each year and not suited for cultivation. Better-drained soils such as Oldsmar fine sand, Wabasso fine sand, and Fort Drum high sands that are characteristic of more northerly private lands are suitable for pasture, citrus, and vegetable crops. These soils support some of the most productive agricultural land in the world. And where interconnected forestland remains in these more fertile northern areas, the land supports both crops and panthers. White-tailed deer were important to panthers throughout south Florida, but as Walt had learned in his studies, they were more abundant on the better soils to the north.

Scats from throughout southwest Florida indicated that hogs were the most commonly eaten food item followed by deer, raccoon, and nine-banded armadillo. When the size and weight of prey are considered, deer and hogs accounted for 86 percent of food consumed in the north and 66 percent in the south. Rabbits, primarily cottontails, were found in similar proportions in both areas and were not important panther foods. Rodents and reptiles were taken even less frequently.

A panther's daily energy expenditure ranges from about 1400 kilocalories for a six-month-old kitten to almost 5000 kilocalories for an adult male. Females raising two kittens require nearly 5500 kilocalories a day to take care of themselves and their growing litter. Based on these energy demands, panthers on average need to kill the equivalent of one adult deer a week. Humans, in contrast, require from 650 to 3000 kilocalories a day from food sources that can vary tremendously. Lactating women require about 2700 kilocalories a day, men require about 2900, while children require from 1300 to 2000 a day.

Panthers that feed primarily on deer should be able to tolerate longer intervals between kills and thus make fewer kills annually than panthers that feed mostly on small prey. Because hogs are smaller than deer, a higher kill frequency is necessary for panthers specializing on this naturalized species. On the other hand, hogs are very easy to catch—in his speedier days I watched Jayde run down hogs in open fields and catch them by hand—so panthers probably conserved energy when specializing on this slower prey item. Panthers residing in the Fakahatchee Strand and other areas with low deer and hog densities must kill at rates exceeding those of panthers feeding mostly on large prey because so many of their kills consist of raccoons, armadillos, and other small animals.

Abundant large prey may not be a prerequisite to healthy panthers. After all, J. L. Yanez and his colleagues in 1986 found that the diets of Chilean cougars were nearly 80 percent European hare and other small prey species. In this case, hares were apparently so abundant that more frequent hunting forays were probably rewarded with a successful meal. If raccoons and armadillos were more abundant in south Florida than they are today, deer and hogs would probably be of lesser importance. Nonetheless, panthers are more productive and have smaller home ranges where large prey are abundant.

The Florida panther is not unusual among its cougar relatives in depending on the most locally abundant prey. These species vary from mule deer and elk in western North America to guanacos and Brocket deer in South America. Only in Chile have small prey dominated the diet of cougars. Domestic livestock were unimportant to Florida panthers, although cougars in the western United States kill thousands of sheep and cattle annually. Apparently, abundant deer and hogs in areas where panthers and cattle overlap prevented the development of a depredation problem in south Florida. I know of no rancher in Florida who has implicated panthers as regular killers of cattle. Panthers appear to prefer wild prey species over larger domestic animals.

Apart from domesticated animals, south Florida's backwoods are also home to a variety of wildlife species that were intentionally or accidentally introduced by humans. Two of these exotics—hogs and armadillos—accounted for 48 percent of the prey consumed by panthers. Based on prey size and weight, exotic wildlife accounted for about 60 percent of their diet north of I-75 compared to about 30 percent in the south. In Florida, both hogs and nine-banded armadillos have relatively high reproductive rates and are easily captured prey. The dominance of hogs in the diet of some panthers living where deer are also abundant suggests a preference for this species. Hogs are most abundant on private land where they are viewed as a game species that can become an agricultural pest. Their habit of rooting and tilling can damage improved pastures, and their fondness for farm produce creates little goodwill among farmers. Although the Florida Game and Fresh Water Fish Commission views the hog as a valuable game species, the state and federal park services usually seek ways of reducing their numbers. Despite the hog's importance to panthers and its potential to damage crops and ecosystems, no recent studies have been conducted on this species in south Florida to better understand its biology and its relationship to panthers.

* * *

South Florida is unique in eastern North America as the only place supporting resident panthers, bobcats, and black bears. And over the last decade, a new carnivore, the coyote, has invaded this part of the state following a half century of range expansion in Florida. Because black bears are predominantly vegetarian, there is little reason to suspect that any meaningful competition exists with panthers. Their occasional consumption of vertebrates most likely results from scavenging or opportunism. Bobcats, however, regularly kill deer and were the most important cause of mortality among the collared white-tailed does we would study in Bear Island. Statewide, bobcats eat mostly rabbits and rodents. But under the unique conditions prevalent in south Florida, deer appeared to be more important to this smaller cat. Water or a combination of water and dense herbaceous vegetation, as in Bear Island and Everglades National Park, may restrict deer movements sufficiently to increase their vulnerability to bobcats—just as deep snow may increase deer vulnerability to bobcats in New England.

Gary Koehler and Maurice Hornocker observed competition among three carnivores in Idaho. They found that competition was avoided through differential prey and habitat use. But when food resources grew scarce, mountain lions were quick to kill their competitors, coyotes and bobcats. In southwest Florida, bobcat remains have not been found in panther scats— suggesting that if competition does exist between Florida's two native cats it has not reached the point of aggression. We did, however, find one panther scat that contained the remains of a bear cub. In Everglades National Park, panthers have been known to kill bobcats, otters, and alligators, indicating that prey resources may be limited in this unforested landscape. With the exception of the Everglades, the abundance and variety of prey, as well as behavioral and environmental partitions (little overlap in food habits, different habitat requirements, and the like), have allowed panthers to coexist with other predators even in areas where their home ranges overlap.

The coyote's potential impact on the panther and south Florida's other native carnivores is anyone's guess. But as south Florida becomes drier and less forested, conditions will improve for this rangeland predator. Coyotes are amazingly tenacious, even when they compete with larger predators for food and space. A large, widespread population of coyotes in south Florida may well reduce the benefits that the well-drained, productive uplands on private lands provide panthers today. In the late 1980s we began seeing coyote tracks throughout panther habitat on many of the ranches north of Interstate 75. Today ranchers and farmers see coyotes trotting through cattle pastures and walking the edges of farm fields throughout Collier and Hendry counties. And coyotes are known to raise their young in the same palmetto thickets that panthers prefer. In 1988, male Panther 24 shared his extensive home range with coyotes that utilized the edges of vegetable farms, citrus groves, and remnant forests in Highlands County. Unlike the equivocal reports of panthers from throughout Florida, the increasing discoveries of both coyotes and their sign suggest that this adaptable creature is on the rise in south Florida. It appears that the landscape changes which have reduced panther numbers have encouraged the coyote's firm establishment in a previously wetter and more forested south Florida. The coyote may already be as well established in south Florida as the wild hog was 500 years ago. And, as in the case of the wild hog, there have been no efforts to understand the importance of this new carnivore on the agricultural community, south Florida's ecosystems, or the Florida panther.

Such could not be said about the white-tailed deer in south Florida. This popular game species had already been the subject of exhaustive ecological monographs published by the Florida Game and Fresh Water Fish Commission during the days when nothing was known about Florida panthers. And we spent an additional six or seven years contributing to this already considerable body of knowledge. We were primarily concerned with the ability of South Florida deer to provide adequate nutrition for panthers. But there were also political dimensions to the study that were the result of conflict between state and federal land managers over the way deer should be hunted—or if they should be hunted at all. While we may not have succeeded in settling the political differences, we did make the most of our efforts to understand white-tailed deer in good panther habitat. Much of our education in this aspect of research derived from a particularly exhilarating activity.

30 May 1987 ■ We were oblivious to the rhythmic thumping of rotor blades as we skimmed the treetops at the edge of the Okaloacoochee Slough in Bear Island. Walt and I had already bagged two deer for Darrell and wildlife pathologist Scott Wright to examine as we hunted for more. We had reached the end of the morning's session as well as the fuel capacity of the helicopter and were making one last pass before heading back to Naples. With both rear doors removed, the view of forests and marsh was spectacular and the wind rushed between our ears and crash helmets. Walt was perched at the right door, fighting back his nausea, with his feet on the landing skids. He wore a steel-buckled, nylon safety harness that secured him to the helicopter. Gripped with both hands and resting tensely on his lap was a weapon straight out of the films *Mad Max* and *Road Warriors*. The gun was based on the frame of a .30 caliber rifle, but instead of one barrel it had four. The firing chamber had been modified to vent the explosive gases that would propel four steel weights anchored at the corners of a 10-foot-square nylon net.

The purpose behind our sorties was to capture and radio-collar a large sample of white-tailed deer in order to understand key aspects of their natural history and behavior. As an important food source for the panther, a prized quarry for sportsmen, and the focus of a never-ending management argument between the GFC and the National Park Service, white-tailed deer had been the target of close scrutiny and speculation for many years. This was the first time, however, that deer had been captured in southwest Florida for radiotelemetry studies. They would be an adjunct, too, to an ongoing study that had examined over eighty deer carcasses from the Big Cypress National Preserve. While the scientific collection of deer with bullets was a somewhat brutal form of research, it was the most efficient way to reveal landscape-scale patterns of their productivity and physical condition. Ultimately our data would reveal that, in general, deer north of Interstate 75 were larger, in better physical condition, and produced more fawns than those of the cypress-dominated swamps to the south—more evidence that the private lands to the north had the greatest potential to support panthers.

But these findings were insufficient to convince agency decision makers that the abundance of panther food was more a reflection of the natural productivity of the landscape than a product of a particular management philosophy. In the Big Cypress National Preserve, where the land was theoretically managed with natural resources in mind, the two agencies responsible for its use (the GFC and the National Park Service) could not agree on the deer's role in the scheme of human use. Although there was no indication that hunters and panthers competed for deer, Bear Island became the site of expanded studies to explore the impact of panthers and human hunters on the white-tailed deer. This newly initiated deer telemetry study would take at least three years to complete.

Seated to the left of Walt, I looked down on a watery landscape while scanning my field of vision for our reddish-brown targets. At lower altitudes, below the height of the tree canopy, the helicopter maneuvered to within a rotor's width of slash pines and red maples. Rounding the corner of a narrow cypress strand, I spotted two does forging through chest-high pickerel weed and maiden cane. I hailed Walt and the pilot over the ship's intercom, pointed to the targets, then waited for the chase to begin. Now it would be up to the pilot to position the aircraft about 15 feet above the ground, just behind and to the left of the fleeing deer, and it would be up to Walt to lob the net in front of one of them. If the shot was perfect, the net would entangle long, dainty legs, leaving the doe helpless while one of us approached to restrain and untangle her. The beauty of the technique rested in the fact that anesthetizing drugs were unnecessary. Several months earlier we were capturing deer at night from the decks of speeding airboats equipped with powerful spotlights and dartguns. We wound up killing two of the four that we captured in this manner. This Spotlight and Dartgun Technique was retired almost as quickly as the Texas Ax Method for removing panthers from trees.

The pilot banked the helicopter sharply to the left and put Walt in position to see his quarry. I craned my neck to watch our approach and the net's trajectory. With the deer lunging through deep water and frequently changing course, Walt experienced a rare miss. Because our spare net had been wasted on our last chase, this miss posed a problem. Not wanting to give up the opportunity to add another adult female to the sample, I asked the pilot to circle the deer and keep her in deep water. We would use another strategy to catch her as long as the fuel held out. Deep water would tire her quickly and I hoped we might have the chance to catch her without the net. Within a few short minutes the deer plunged into a clump of sagittaria and sawgrass, standing motionless, with just her head and laid-back ears above water. Skillfully our pilot positioned the helicopter about 20 yards behind the deer and about a foot above the marsh. The rotors pounded air into the marsh creating a circle of white water where it normally flowed at a snail's pace. While Walt unbuckled himself from the seat, I bailed out of the ship.

I now found myself in waist-deep water. The roar of the helicopter had certainly overwhelmed the senses of the exhausted deer, for I was able to wade within arm's length. But the waves I was making alerted the deer to my presence and she once again attempted her escape. As she surged forward I leaned and grabbed for a leg, hoping to avoid a thrashing with sharp hooves. By the time I had a firm grip we were both totally submerged. Her fatigue allowed me to reel her in like a large catfish until I had both arms wrapped around her rib cage and both of my feet back on marsh bottom. At this point she appeared remarkably calm as Walt, only half as wet, arrived to give me a hand. Within ten minutes the helicopter returned with Darrell and Scott Wright, who officially inducted Deer 18 into our study. She was a large doe weighing 85 pounds—as large as any female panther in the area. Although her radio collar failed within a few months after her capture, for several years she was seen regularly during telemetry flights, often accompanied by offspring.

We captured sixty deer using the net gun method—very few managed to avoid the nets draped over them by marksmen Walt and Jayde. As our sample grew and we learned more about each one, the picture of deer life in the marshes of Bear Island began to unfold. The annual mortality rate of adult does was less than 20 percent—suggesting that plenty of females survived each year to reproduce and raise young. Causes of deer deaths included black bears, alligators, bobcats, people, and, of course, panthers. Even so, the deer population in Bear Island was stable and fully capable of compensating for predation, the number one source of mortality. Because Bear Island contained the best panther habitat within the Big Cypress National Preserve, we fully anticipated that panthers would account for the most deaths of radio-collared does.

To our surprise, bobcats killed more than twice the number of does that panthers killed over a span of nearly six years (September 1986 to June 1992). Despite the bobcat's prominence as a killer of an animal that was often four times its size, white-tailed deer were still an insignificant part of the bobcat's diet. According to a scat analysis conducted by Darrell, deer occurred in only 5 percent of Bear Island bobcat scats versus 28 percent of panther scats. Just like bobcats throughout Florida, Bear Island bobcats were rodent and rabbit specialists, but their tolerance for more open habitat allowed them access to deer in areas that panthers typically avoided.

The apparent contradiction between observed doe mortality and bobcat food habits was likely a function of scale. Bobcats use small home ranges and move around less than panthers. Panthers are veritable nomads compared to their smaller kin. Whereas a male panther may spend a few weeks each year in Bear Island, a male bobcat may live his whole life in an area half that size. On an acre-for-acre basis, bobcats not only use Bear Island more intensively than panthers, but they must also be much more intimate with every cypress dome, each buggy trail, and, more than likely, with each doe found within their home ranges. While bobcats went about their day-to-day business in search of cotton rat and marsh rabbit–sized prey, every now and then they were afforded the opportunity to feed on deer.

Some deer appeared to have been killed by bobcats in the same manner as panthers dispatch prey. Others, however, suggested great opportunism. A female and her fawn were killed while the doe was preoccupied with labor. The others were probably killed while at rest on their beds of sawgrass or ferns. Although bobcats might appear to consume deer more frequently than panthers did, this conclusion was contradicted by Darrell's bobcat scat analysis. And our work on panther food habits showed that deer were, indeed, taken frequently by panthers. I suspect that the few bobcats that killed deer in Bear Island simply had many more opportunities to do so than the panthers who passed through now and then. Had panthers chosen to maintain home ranges as small as those of bobcats in Bear Island, their need for large prey items would no doubt have pushed the local deer herd to the brink of collapse. This is an important reason why panthers have such large home ranges—to spread the impact of their predation over time and space and thus allow prey numbers to sustain a level of natural mortality that is kept in balance with reproduction.

Compared to bobcats and panthers, deer have tiny home ranges. Seventeen adult does used 480 acres on average, while two bucks used 1166 acres and 2591 acres, respectively. Does were very consistent in their movements, which usually took them to forested uplands such as hardwood hammocks and pine flatwoods. Even so, they were also seen frequently using the same marshes in which they were captured. Because most deer maintained home ranges with diverse plant communities, large seasonal shifts were uncommon. As little as a few hundred feet might separate succulent marsh grasses, lilies, and willows from fruit-producing upland plants such as live oak, gallberry, and saw palmetto. Neither food nor cover seemed to be in short supply at any time of the year for deer in Bear Island. And this population appeared to provide ample resources for the many species that depended on it in some fashion or another. The combined effects of predation by alligators, bears, bobcats, panthers, and even people were fully absorbed by the white-tailed deer living in this diverse corner of Florida panther range.

Wild hogs were a different matter. As an introduced exotic, they are disdained by many landowners, both private and public, because of the damage they can inflict on crops, pastures, and natural environments. At the same time, hogs were, as we had learned, the primary food item for panthers, especially on the more productive private lands. Perhaps the anxiety with which wildlife agencies view private landowners explains the difficulty I had in convincing agency administrators that we needed to know more about the wild hog and its role as a component of panther habitat. Florida was likely the point of the first North American introduction for the species when Hernando de Soto landed in 1539 near Charlotte Harbor. Today hogs reach peak densities of more than ten per square mile on well-drained upland habitats within the current range of the panther. They have probably been important panther prey for more than 450 years.

What little can be said about the biology of wild hogs in south Florida must be inferred from other parts of the species' range. Hogs are notorious for maturing quickly (five to ten months) and reproducing frequently. Sows can have as many as a dozen young every year—and, moreover, are capable of breeding throughout the year. Boars are solitary, except when females are in estrus, and females are often seen in large groups with other females and their young. Throughout most of the hog's range, hunters cause more mortality than any other known factor and predation is considered insignificant. But the effects of predation on hogs in southwest Florida—by panthers, certainly, but also by black bears, bobcats, and coyotes—may be very different from what has been observed in other parts of its range.

There are too many unanswered questions about this prey species. Here is an animal that may have been responsible for the panther's survival through a period of deer eradication and the clearing of much of the panther's preferred upland forests. And yet, because of its prevalence as a carrier of pseudorabies, Melody Roelke believes it may have prevented the return of the panther into the productive ranchlands of Glades, Charlotte, and Highlands counties. While this notion is supported only by the evidence of one panther death due to this virus (Number 29) where panthers are doing well, pseudorabies might have been responsible for prematurely snuffing out the life of male Panther 24 in Highlands County.

Unfortunately, we will never know. Lack of interest in panther-related hog research has left us totally in the dark about the true status of this centuries-old relationship. We have no idea how hogs respond to floods and droughts. We do not understand hog mortality and recruitment patterns— prerequisites for gauging the impacts of harvest and predation on the population's ability to increase. In essence, without a basic understanding of the wild hog's ecology in southwest Florida, management agencies know next to nothing about this species' impact on the panther's health, nutrition, survival, and distribution. Perhaps the idea of hog management seems trivial to agencies that are already overwhelmed with other political obligations.

THE LURE OF CAPTIVE BREEDING

Panther 19 was a prime example of one of the most commonly cited symptoms of panther endangerment: as an adult she engaged in incest, producing several litters that were sired by her father, Male 12. One of the signs of such inbreeding is diminished reproductive output, at least in theory, and our early experiences in the field did little to suggest that this was not already happening among Florida panthers. From 1981 through 1985, we handled no kittens—and this, we assumed, stemmed from reproductive problems in the population. Thus Female 19 was very important in our emerging understanding of panther ecology. She demonstrated to us several firsts, not only for her endangered population but for cougars in general. Although she represented only a sample of one, she forced us to pause and question our assumptions.

15 August 1988 ▪ Jayde and I were making a routine ground inspection of Panthers 11 and 19, hoping to turn up a track, a scat, or a kill. There was nothing special about their locations except that their proximity to each other on the small Sunniland Ranch in Collier County would allow us to see what both of them were up to. We had found Female 11 resting in a labyrinth of limestone solution holes with a lush canopy of tropical shrubs, ferns, and vines. The exposed rock and dense cover made this a particularly cool spot in the oppressive sweltering weather, and the observation reinforced our impression that such protected sites were key components of panther habitat. Number 11 had slinked away at our approach, so after noting the details of her rest site we continued on our way through live oaks and fern-studded cabbage palms toward her daughter. As we neared Female 19's radio signal, Jayde and I split up to cover more ground in case she had stashed a kill nearby. Because she had not yet left her mother's home range, we wondered if Female 11 still tolerated her daughter at kills. At nearly two years of age, Female 19 would be dispersing soon and making room for Female 11 to have her next litter. As the first female kitten to be radio-collared, Female 19 would help balance our male-biased statistics of panther dispersal. In addition, we hoped she would survive to adulthood and produce kittens of her own.

My route took me through open palmetto thickets shaded with live oaks and Spanish moss. Number 19's radio signal remained constant as I approached a drought-stricken pop-ash slough that, during the years before widespread drainage, would have held knee-deep water. Today the slough's only contents were cool, moist soils covered with parched, decaying leaves. Normally I would have stayed out of this depression to avoid getting wet and making noise, but the slough was a dry path of least resistance. Now her signal was thumping from the receiver. Surely I was close enough for her to hear me.

Based on previous experience, I assumed that Female 19 would be resting in the palmetto thicket just beyond the slough. It was as thick and dense as any daytime rest site we had noted in previous walk-ins. As I scanned the squat, buttressed trunks and fallen branches for the easiest path forward, my eyes were drawn 20 feet away to a furry mound at the base of one of the pop-ash trees. It was Female 19. Immediately, I assumed something was wrong. We had never surprised a panther in the dozens of times we had walked in on them—their keen senses were like radar. And seeing a panther lying in a fairly open setting such as this was strange to say the least. At this point, events happened rapidly. I realized that Female 19 was not alone when a four-month-old kitten sat up and looked me in the eye. It yawned, shook off a few mosquitoes, then lay down next to Female 19's feet. As my focus sharpened, I saw three other spotted mounds napping by the mother's head. All the while Female 19 lay peacefully beneath the tree. The sound of a branch breaking underfoot and the swishing of palmetto fronds finally got her attention. She quickly sat up, looked directly at Jayde (who was now about 50 yards away), then turned her head toward me. Her amber eyes were like glowering stones in an expressionless feline sculpture. A low guttural growl sent the kittens scrambling for the palmettos. And then she began a series of slow, deliberate steps in my direction. With only the receiver and antenna for defense, I hoped that Jayde could rescue his unfortunate supervisor before it was too late.

But Jayde would not have to demonstrate his loyalty after all. Apparently my slow retreat showed her that I was not a threat, so she halted her advance and allowed me to escape unharmed. In the span of a few heart-pounding moments, Female 19 taught us that Florida panthers do not always follow the stereotypes described by previous studies of mountain lions. In an instant, the established norm for the species had been revised. Mating with Male 13 five months before her second birthday (an event now understood in retrospect) not only represented the earliest known age of first reproduction for a female Florida panther. It was the earliest published record for the entire species. And Numbers 11 and 12 became the first documented panther grandparents.

Very little changed for Female 19 over the next five years—she continued to be very busy. With the exception of a home range shift, forsaking her mother's territory for what would become the Florida Panther National Wildlife Refuge, she remained a predictable study animal. She also continued to be quite productive. Including her first litter of four, by the winter of 1993 she had produced at least ten kittens from four litters. This was equivalent to producing two kittens every year during her adult years as a study animal. The average for the species is about one kitten per year.

* * *

The reproductive performance of females like Number 19 was a part of modern panther natural history that never made the headlines. High kitten production more than offset adult mortality. And, better yet, we continued to capture young, independent panthers that were the products of uncollared parents. Often they appeared out of nowhere.

Mortalities have always been the events driving panther headlines, even though adult female mortality was unusual. Three of the four that died between 1985 and 1993 (Females 8, 18, and 41) were not significant losses to the population because they had not produced a single kitten during a combined total of eight years of monitoring. The other female mortality statistic was a panther run over on Alligator Alley in 1986. Although she was uncollared and had an unknown reproductive history, a necropsy suggested that she had in fact given birth to kittens in the recent past. Eight deaths were independent subadult males or dependent kittens, two were dependent female kittens, and nine were adult males. Of these adult males only four (Males 13, 17, 20, and 37) were known breeding adults and only two (Males 17 and 37) were resident adults. The cause of death was discernible for twenty of these panthers: five were hit by cars and ten (50 percent) were killed by adult male panthers. The other causes of death included rabies (Number 33), pseudorabies (Number 29), a bacterial infection (Number 35), old age (Number 8), and congestive heart disease that was probably exacerbated by Number 20's stay in captivity and his badly worn canines. Clearly panthers themselves could be their own worst enemy. But they more than compensated for these losses with ample reproduction.

Yet it was the dead panthers that made the headlines throughout Florida for many years. In the early days (1980 to 1986), the newsmakers were panthers struck and killed on Alligator Alley and State Road 29. In the public's mind, each death resulted in a net loss from the "official" population estimate of thirty to fifty panthers. It was just a matter of time, therefore, before a speeding beer truck or Winnebago flattened the last of the breed. What too few people knew, despite our reports and publications, was that panthers, like most normal populations of wildlife, reproduced. And as far as our sample was concerned, kitten production was outpacing adult mortality. But somehow our data and reports never seemed to get past midlevel agency personnel.

An increasing number of captures after 1986 not only improved our understanding of panther habitat but gave us a more accurate view of mortality patterns. There was more to panther deaths than chrome bumpers and concrete. As the early days of panther research in the Fakahatchee Strand and other marginal parts of the panther's range began to recede, it became clear that the old assumptions about the panther were wrong. Female 19's remarkable reproductive success turned out to be not the exception but the rule. Stability in resident adult home ranges encouraged regular reproduction, but it also led to the bulk of mortality that occurred among young males. Nevertheless, there did not appear to be any interest in acknowledging that roadkills might not represent simple subtractions from the total. Our new data did little to alter public and agency perceptions. Perhaps the agencies responsible for management and research were simply reluctant to give credence to information that seemed to contradict the detailed plans that had stemmed from the first gatherings of the Florida Panther Recovery Team.

Female 19 was just one element of the panther maternity ward in southwest Florida. From 1985 to 1994 we studied fifteen adult female panthers, and all but four produced kittens during this time. We monitored each of these females for a minimum of seven months to over ten years. Number 9, the record-holder for years tracked, produced at least four kittens over an eight-year period, Number 11 produced at least eight over seven years, Number 31 produced seven kittens in four years, Number 32 produced three in four years, Number 36 produced four in four years, Number 40 produced seven in three years, and Numbers 48, 49, and 52 all had their first litters in 1992 or 1993. We documented the age and sex of thirty-six of these kittens by capturing them with hounds, handling them as neonates at their dens, or discovering them as roadkills. The presence of others was verified by field sign left by families as they traveled together. Of the three females we handled as kittens, every one of them had reached adulthood and produced their first litters before they were two years old. While this vigorous productivity was an encouraging sign for panther recovery, it also came with an equal dose of frustration. Only 11 percent of the kittens we identified over the course of eight years survived to become resident breeding adults in the wilds of southwest Florida. No male offspring of collared panthers became residents between 1985 and 1994. And herein was hidden the rest of the story—the flip side of the rosy picture of panther obstetrics.

Despite Female 19's precocity and reproductive vigor, only one of her kittens (Number 45) is known to have survived beyond three years of age. Of those we captured, Number 30 was killed by Number 37, Number 43 was killed by Number 26, an uncollared male killed Number 53 (his sister was killed by a car), and Number 205 was taken into captivity. Nonetheless, after nearly a decade of research, the Florida panther gave no indication of succumbing to the pressures we so freely discussed at meeting after meeting. Yes, panthers were dying. But high kitten production was more than offsetting the infrequent mortality of resident adults.

*   *   *

As we rewrote the natural history of the Florida panther, the heads of government agencies were at a loss for saving the subspecies. Panthers were viewed as a lost cause, the public was criticizing agency efforts, and no clear progress had been made toward implementing the panther recovery plan. As a result, despite a wash of good news from the field, agency anxieties led to substantial changes in strategy that threatened to overwhelm the panther. But Female 19 and her kin were oblivious to the machinations of their self-appointed caretakers and simply went about their long-evolved business of death and renewal.

The lure of panthers in captivity had been a compelling attraction since the inception of modern panther work in the 1970s. If the population was indeed as small as it had first appeared, then it would not be long before zoos became their last hope. Soon the topic of captive breeding was receiving more publicity and debate than any other activity related to panther recovery. Indeed, I might be remembered as one of its more ardent opponents. This, however, is an unfortunate (and inaccurate) portrayal of my personal philosophy. I began work on the project in 1985 knowing full well that we were working toward the establishment of a captive population to be derived from the wild. I acknowledged this without reservation and understood the steps outlined in the Florida Panther Recovery Plan. Basically, experiments with Texas cougars in captivity and in suitable reintroduction sites would help us develop handling procedures that would ensure smooth sailing when the day of reckoning arrived.

The panther's recovery plan seemed to be a sane and reasonable document.

Essentially the plan called for conducting telemetry studies of panthers in south Florida while conducting experiments with Texas cougars in north Florida. These experiments were designed to examine the potential of currently unoccupied range as future panther habitat and to develop handling and conditioning procedures for captive-raised panthers. The methods perfected on a nonendangered subspecies, the Texas cougar, would then be used with Florida panthers in captivity and when suitable habitat was found for reestablishing panthers outside south Florida. The five years it would take to reach the point where panthers would leave the wild for captivity would surely afford us the chance to really understand panther biology—if enough of them still existed to study.

For some, however, no plan is swift enough. A few midlevel administrators in the Florida Game and Fresh Water Fish Commission and the U.S. Fish and Wildlife Service succeeded in circumventing the process that had been worked out in the late 1970s and early 1980s. The experimental approach called for in the recovery plan was shelved—based on the prediction of a mathematical population extinction model developed by Ulysses S. Seal and his colleagues with the Captive Breeding Specialists Group. This computer modeling was a service they had also provided to those working on the black-footed ferret, Javan rhinoceros, and Puerto Rican parrot. While all of these endangered species were similar in exhibiting low numbers, much more was known about them than we knew about panthers.

Even the most respected wildlife population modelers admonish anyone who uses a computer model for making significant management decisions. Models are extremely useful in illustrating our depth of understanding of a population of interest and in identifying gaps or weaknesses in data. They are best used to help researchers focus on key biological questions, not management actions. And at this point in panther recovery, there were fewer strengths than weaknesses in our understanding of the panther's natural history. Nonetheless, by utilizing data collected mostly from 1981 to 1985 (when the panther was characterized as an old, anemic, disease-ridden, and nonreproductive population), the model developed by the Captive Breeding Specialists Group helped solidify predictions of the big cat's imminent extinction. The scenario painted in their 1989 Florida Panther Viability Analysis and Species Survival Plan was not optimistic: a small population with rapidly dying, inbred adults. Moreover, the species survival plan predicted a population decline of 6 to 10 percent a year coupled with a 3 to 7 percent loss of genetic diversity with each generation. Little wonder, then, based on this dire vision of the future, that captive breeding was viewed by many as the only solution to the panther's seemingly insurmountable problems. As many panthers as possible—and as soon as possible—would have to be removed from the wild in order to stop this inexorable spiral into oblivion.

This was a disturbing recommendation for two reasons. First, the removal of all known individuals would permanently prevent us from characterizing the true status of the population. Our tiny radio-collared sample was the only window into the relationship between panthers and their landscape—and these would be the individuals targeted for removal. While the Captive Breeding Specialists Group assured us that captive panthers would garner more public support for all aspects of panther recovery, those of us in the field knew that the absence of free-ranging animals would eliminate another reason to fight for land-use reform and expand efforts to acquire more panther range. And second, the data used to drive the model were suspect. Using kitten mortality estimates based on conjecture, as well as population figures from panthers living south of I-75, the model characterized a population that did not exist. (Only eight, mostly old, and predominantly male panthers had been collared and tracked between 1981 and 1985.) In any population model, high kitten mortality, low adult survival, and a population skewed toward males inevitably result in decline.

At best, then, the model represented only a segment of the population, mostly in the Fakahatchee Strand, during a short period of time—a segment that to this day acts as a population sink, absorbing more reproduction from elsewhere than is produced locally. Our capture successes had painted a very different picture of panther population dynamics. The numbers driving the model and the numbers representing the reality we observed were quite different. The Vortex model used in the exercise was based on certain assumptions: a maximum population of forty-five, age of first female reproduction of thirty-six months, kitten mortality of 50 percent, annual adult mortality of 25 percent, and population demographics that were unstable. My calculations suggested a population of at least seventy, age of first female reproduction of eighteen months, kitten mortality less than 20 percent, annual adult mortality less than 20 percent, and population demographics that were stable. In fact, the discrepancies between these figures were the difference between a population that was in trouble versus one that had the potential to grow.

*　*　*

One of my objections to the plan developed by the Captive Breeding Specialists Group, therefore, was its misrepresentation of panther demographics and the use of unreasonable or incorrect variables in its extinction model. I expressed these shortcomings in written comments and sent them into our agency's convoluted chain of command on 29 May 1989. Even in a population model as well designed and powerful as this one, high mortality estimates inevitably force the calculations to predict rapid extinction. I disagreed with their figure for genetic erosion, too. More adults were successfully breeding than the few that had been collared so far, we argued. We concluded: "The south Florida population [of panthers] is too valuable to experiment with the consequences of untested and possibly inappropriate manipulations."

My written statement, as well as innumerable comments at meetings, went unanswered. Shortly thereafter the Florida panther species survival plan was accepted by the Florida Panther Interagency Committee. Panthers were then slated for removal. While most of us working with panthers in the field did not agree with the steps being taken, there was little sense in fighting the process—especially when no one appeared to be listening. Halfway through the 1990 capture season, the last before removals were scheduled, I drafted another memo to recommend a safer approach to establishing a captive population. In this correspondence, I noted the experiences of veteran curatorial staff from the Miami Zoo as well as those of past Panther Recovery Team member and feline breeding expert Robert Baudy. These authorities urged that adult panthers should not be targets for removal because they seldom adapted to captivity well enough to breed. Kittens, they independently agreed, were a much better risk for removal—and besides, very young cats were much easier to replace by natural reproduction than adults. Moreover, I observed, regular removals of adults from the wild would just add to mortality and the population might spiral into an irreversible decline. I recommended that only kittens be removed.

This correspondence did not go unanswered. My letter, intended for GFC consideration, was forwarded to the general curator of the Miami Zoo, Bill Zeigler. Several other Florida zoos, including the Jacksonville Zoo and the Lowery Park Zoo in Tampa, were very interested in obtaining Florida panthers for captive breeding. After all, this is what zoological parks are good at—and housing such rare specimens would certainly attract more visitors and help cover the great expense of housing large cats. (There were no plans to subsidize the zoos for their assistance.) Even though my letter might outline some serious concerns, I was told, in correspondence from Mr. Zeigler to Mr. Tom Logan, that I was insufficiently knowledgeable to have an opinion on the subject.

On 15 February 1990, the U.S. Fish and Wildlife Service published a notice in the *Federal Register* "to issue Endangered Species Permits for capturing Florida panthers from the wild population in order to accelerate the establishment of a captive population." The removal plan called for the establishment of "500 breeding adults in captivity." The FWS proposed the removal of twenty-five animals over three years—seven adults and eighteen kittens. While offspring produced through the breeding program "could provide animals which will be needed in re-establishing additional populations," the notice went on to state that the reintroduction "aspect of the recovery effort would be speculative since the ability of a captive breeding program to produce a sufficient number of animals for a reintroduction program is unknown at this time. Such a program, if warranted, would be a separate action." Despite the lack of protocols for managing a captive population and protecting panther habitat, momentum was growing to ignore the cautious approach outlined in the recovery plan. In June 1990, the FWS followed its *Federal Register* announcement with a draft Environmental Assessment that outlined a preferred alternative to remove up to fifty panthers from the wild—including up to six kittens per year, four adults the first year, and one pair of adults per year for the following five years. (Even the mention of panthers as "pairs" suggested ignorance of basic panther social structure and the polygamy that prevails in the wild.) The unspoken effect of these removals would be simple: the elimination of all known breeding panthers in the wild.

My staff and I were not the only ones who were uneasy with the plan. Others, elsewhere, were working to force the government into a more responsible approach to panther recovery. Chief among these citizens was Holly Jensen, an outspoken animal rights activist from Gainesville. Known mostly for her opposition to the use of animals in medical research, she opposed the removal of panthers primarily because there were no provisions for protecting important habitat in south Florida. Jensen expressed her anxieties at many agency meetings where she must have noticed the polite indifference afforded her by most officials. In addition to Jensen, the Fund for Animals had asked the State of Florida and federal government to look more closely at the potential impact of removals on the south Florida panther population and its habitat. This animal rights group, based in Washington, D.C., requested that the FWS prepare an Environmental Impact Statement (EIS) that not only reviewed removal impacts but also examined reintroduction and habitat issues to ensure there would be secure reintroduction sites for the hundreds of panthers that the FWS claimed would be bred in captivity.

On 19 December 1990, the FWS issued a one-page "Finding of No Significant Impact" (FONSI) in the *Federal Register* as well as a notice that the final Environmental Assessment was now available. The assessment contained no examination of habitat that would be used to reestablish panthers. Nor did it analyze the threats to panther habitat, even though residential development, agriculture, and other human activities were the roots of the panther's problem. Nonetheless, removals were scheduled to begin in less than a month. In response to what was perceived as inadequate review, the Fund for Animals and Jensen filed a lawsuit on 15 January 1991 under the National Environmental Policy Act (NEPA) and the Administrative Procedure Act requesting declaratory and injunctive relief against John Turner, director of the U.S. Fish and Wildlife Service, and Manuel Lujan, secretary of the Department of the Interior. Such lawsuits are remarkable in their ability to galvanize government agencies.

The NEPA review process—in particular, the preparation of an EIS—has two purposes. First, it requires an agency to review the impacts of a proposed action and "ensures that important effects will not be overlooked or underestimated only to be discovered after resources have been committed or the die otherwise cast." But, second, NEPA also serves a larger informational role. It assures the public that the agency has indeed considered environmental concerns in its decision-making process—and, perhaps more significantly, provides a springboard for public comment. Actions requiring an EIS included the construction of I-75 through south Florida, the reintroduction of wolves into Yellowstone National Park, and the general management plan for Big Cypress National Preserve. To Holly Jensen and her colleagues, the removal of panthers for captive breeding was certainly an action equivalent in impact.

The FWS, however, refused to prepare an EIS on any aspect of the panther captive breeding proposal. In their lawsuit the Fund for Animals and Jensen argued that the Environmental Assessment examined only the impact of removing panthers from the wild even though the FWS was required to prepare an EIS that reviewed the *entire* removal, breeding, and reintroduction program for panthers. NEPA requires that an EIS for the whole program must be prepared if proposed actions are "connected," "cumulative," or sufficiently "similar" that a comprehensive EIS is the best way to identify the environmental effects. In this case, the capture, captive breeding, and reintroduction of panthers were the connected actions. Even the Environmental Assessment noted this relationship, stating that "the proposed captive program would consist of a number of interrelated actions and segments." Basically, they would consist of the animal selection process, the capture, captive management, captive breeding, conditioning, reintroduction, and reestablishment of the population. Nonetheless, reintroduction was not addressed to any meaningful extent. If reintroduction issues were not addressed when the scope of the program was being decided, rather than at the reintroduction phase as the FWS wanted, crucial actions might be sidestepped that would otherwise ensure that adequate habitat was available when the time came to release the panthers.

Moreover, the plaintiffs argued that the FWS had failed to prepare a specific EIS on the impact of removing fifty animals, including adults, from a population that was officially estimated at the time to number between thirty and fifty animals. While the final Environmental Assessment stated that the FWS would only remove panthers that were not "contributing reproductively to the wild population," the draft assessment had said there did "not appear to be any non-breeding adults in the wild population at the present time." This raised concerns that the FWS, in order to remove the number of adults specified in the Environmental Assessment, would in fact permit the removal of breeding adults.

In many open meetings I had argued that removal of resident adults would disturb the panther's social structure, but my warnings, like my memos, had no discernible effect. Darrell, Walt, Jayde, Roy, and I were by now no more than sideline spectators in an increasingly vigorous contest. Our advocacy could be for nothing more than interpreting and presenting the facts as we saw them. Losses of resident adults could lead to vacant home ranges, increased aggressive interactions among younger males vying for breeding rights, and local interruptions to kitten production. Similar problems had been documented in tigers by Mel Sunquist and his colleagues in Nepal, where the loss of a single dominant male resulted in a cessation of breeding that lasted for more than a year.

The nature of the problem was already hinted at in south Florida after the unexplained death of resident Male 17 in southern Hendry County on 23 July 1990. Following this death of the largest panther on record (154 pounds in 1989), adjacent Males 26 and 28 began a series of home range shifts that eventually filled Male 17's home range and brought them into closer contact with each other. Number 28 was the newest arrival in the area, and shortly after Male 17's death he killed adult Females 41 and 18 over the span of one week—the only documented deaths of resident adult females due to male aggression in thirteen years of study. Number 28 himself was eventually killed as the result of a fight with Male 26 in late September 1992. I suspect that the social upheaval which lasted almost two years was caused by Male 17's disappearance. It was this, I think, that motivated Male 28's unusual behavior. He may have viewed the two females not as solitary females without kittens (that is, as potential mates) but as competitors in an unfamiliar territory. Such intergender mortality is not unprecedented among cougar populations. At any rate, Male 26 became a stabilizing influence in Male 17's old home range and reproduction continued without measurable interruption—but with two fewer adult females for the next few years. Clearly, breeding residents were not only essential to continued population productivity, but a surplus of males was needed to fill vacancies when they occurred. The removal of adults for captive breeding—the crux of the panther recovery plan—had the potential to multiply the occasional disruption that occurred naturally.

Holly Jensen and the Fund for Animals also took issue with the FWS's refusal to introduce individuals from other cougar subspecies into the wild Florida panther population. NEPA had been violated, they argued, because public comment was unsolicited regarding the ability of planned genetic introgression to replace captive breeding. The Environmental Assessment supported the need for a captive breeding program on the grounds that "genetic variability within the [wild] population is extremely limited and documented inbreeding is further compromising existing Florida panther viability." But, the Fund for Animals believed that genetic restoration, through the use of related subspecies, could address the lack of genetic variability while at the same time eliminating the need for removing wild panthers. (As we would see later, however, this approach was not without its own problems.) The Environmental Assessment rejected genetic restoration because there were a "number of questions that must be answered . . . before this alternative can be employed." Just like the review of panther removal impacts, it appeared that the FWS was rushing to implement captive breeding without reviewing alternatives.

*   *   *

On 6 February 1991, the parties reached a settlement. The new agreement would allow the capture, care, and captive breeding of up to six panther kittens that were twelve months of age or younger. The agencies also agreed to maintain the kittens in conditions that would allow their return to the wild if the Fish and Wildlife Service decided to end the captive breeding program. While the agreement still short-circuited the original recovery plans, we seem to have communicated our message: adult panthers should be left alone. Moreover, the FWS agreed to prepare a supplemental environmental assessment of the captive breeding program by 30 November 1991. The FWS also agreed that, except for the six kittens, no further permits would be issued until the supplemental assessment was complete. In the event the FWS selected a captive breeding option, the public would have an opportunity to comment on applications for permits to capture or breed panthers. The supplemental assessment was also supposed to provide a thorough analysis of how the removal of adults and kittens would affect the social structure of the population. (To my knowledge, this was never done.) The FWS agreed to "rigorously explore and objectively evaluate" the alternative of genetic enrichment of the panther through the addition of other subspecies of *Felis concolor*. This analysis was to include the environmental, legal, and regulatory impact of the proposed action as well as the genetic enrichment option so the public would have a comparison.

On 29 July 1991, a draft supplemental assessment was released in the *Federal Register*. The assessment elicited this response from the Fund for Animals' attorney, Eric Glitzenstein, in his 11 September 1991 letter to FWS Regional Director Jim Pulliam:

> Tellingly, the Service has not even seen fit to share its "thorough assessment" with the public but, rather, has merely listed factors it "considered" in coming to its conclusion. The Service's implicit

assumption—which plagues its entire approach to implementation of the captive breeding program—is that the public is supposed to simply take it on faith that the agency has in fact engaged in a "thorough assessment" of the enumerated factors and, if it has, that its review of this crucial issue would not benefit from public input, including peer review by such experts as Dr. Lawrence Harris, Dr. [Reed] Noss, Dr. Tony Povilitis, and Dr. Melvin Sunquist—all of whom submitted affidavits in *The Fund for Animals v. Turner,* raising questions about the need for, and wisdom of, removing adults from the wild population.

Both Glitzenstein and Jensen went on to criticize the draft as cursory and inadequate.

The legal contest, it seemed, was not over. The continuation of the lawsuit, while supported by animal rights activists, was condemned by other environmental groups. Some charged that the plaintiffs did not care if the panther became extinct. Sensing, perhaps, this divergence of opinion, a 16 August 1991 Game and Fresh Water Fish Commission press release suggested that the Fund for Animals lawsuit would delay conservation and drive the panther to extinction. But in fact the population was stable. And unlike the critical situation involving the California condor, the FWS had not attempted to invoke the emergency procedures of NEPA, which allowed an agency to proceed without the normal environmental review. Even if the lawsuit delayed the removal program for a year, panthers would still be holding their own.

A final agreement was reached on 5 February 1992. After more than a year of legal wrangling, it looked as though captive breeding would move forward with more than a few changes to the original program. By this time, we would have welcomed any conclusive decision, even if it meant removing some of our study animals. Perhaps the most troubling lesson to emerge from all the turmoil was that the government agencies responsible for panther recovery had to be *forced* to abide by their own rules. But at least the settlement of the lawsuit would give us an opportunity to learn about a critical aspect of panther ecology that we had, as yet, been reluctant to explore.

INTO THE
PANTHER'S DEN

9

Recognizing the critical importance of successful reproduction to the future of the Florida panther, we were hesitant to do anything that might compromise a panther pregnancy or disrupt a panther family. So for five years we monitored reproductive behavior from a safe distance and made educated guesses about the number of kittens in each new litter. The early start of the captive breeding program, however, combined with an emphasis on the young, forced us to increase the number of kittens we handled. From a purely scientific perspective, this was an exciting new challenge that would allow us to compare the Florida panther's productivity with other cougar populations. But our margin of error would become very slim indeed.

During the first year we succeeded in obtaining three kittens of each sex from four different females. They ranged in age from six to ten months and were captured between 20 February and 6 May 1991. In the cases where we removed only one of two kittens, the females temporarily altered their normal movement patterns by making unusually long excursions for about one week—apparently in search of their lost offspring. In the two cases where we removed both kittens, their mothers did not exhibit unusual movements and both rebred within three months. Only one, however, succeeded in raising more kittens. Number 9, living in the mixed swamp forest of the Fakahatchee Strand and by now more than a decade old, had apparently reached the end of her reproductive life. The two kittens we removed were sired by Male 37, who died before he could produce more offspring. Number 9's repeated encounters with Male 51 produced a litter in 1993, but she was unable to raise the kittens past one month of age. Although she was still alive at the end of 1996, she had not produced more kittens.

The capture of dependent but mobile kittens was a complicated business made possible only by the radio collar worn by their mothers. Because the family group was frequently not together, we were successful in treeing a kitten in only about half of our attempts. As we had learned in our extensive telemetry flights, females accompanied their kittens less than 25 percent of the time. The adult female, in her constant efforts to feed her kittens, would often be alone, returning to her offspring after making a kill, or actively seeking her next victim. Given the unpredictability and inefficiency of our strategy, we began to focus on newborn kittens at their den. In this way, we thought, we would cause less disturbance to the family, have greater flexibility in choosing kittens for captivity, and allow the females that had lost entire litters to rebreed at once.

New Mexico cougar biologists Ken Logan and Linda Sweanor told me that young, helpless kittens could be handled safely while their mothers were away. These researchers would approach a den, usually located on a rocky hillside on the White Sands Missile Base, and then wait for the female to leave. Ken and Linda handled dozens of kittens in this manner without incident, and they were rewarded with valuable information on sex ratios and litter sizes. Unfortunately for us, south Florida did not offer boulder-strewn hillsides for panthers to den in. Nor could we gain an elevational advantage from which to observe. In our flat, forested, palmetto-clumped terrain, the panther's landscape of choice, visibility equaled 5 feet—the south Florida analogy to Rocky Mountain hillsides where visibility equaled 5 miles. As a result, we had to rely on our best guess of the den's location in order to find kittens. After determining that the female had given birth to a litter (based on changes in her movement patterns), we would make an approximate location of her den on the ground, home in on her radio signal, and identify helpful landmarks while she was presumably at home. Then the wait was on. A lone cabbage palm or pine snag might provide a visual cue to guide us. Once the female left to hunt, we had a small window of time to launch a search for her den without worrying about a face-to-face encounter with a protective mother. Usually, though, we were forced to make repeated trips before the female left the den to hunt.

6 April 1992 ▪ On this warm, clear spring morning we made our first attempt to find panther kittens and remove them from their den. Darrell, Carolyn Glass (our stand-in veterinarian), Marnie Lamb (her assistant), and I headed out to eastern Collier County to search for Female 40's kittens. We loaded ourselves and our equipment on the swamp buggy, then departed from an old hunting camp built next to a grassy airstrip. We were about halfway to the den site when the steering failed, leaving the two front tires pointing in different directions. Sprawling mostly in mud, I took a rust-welded nut off a suspension spring U-bolt and put the front end back together. By now, however, most of the day was gone, and with it our first opportunity.

7 April 1992 ■ On our second try we had better luck with our equipment (we used a different buggy) and arrived near the panther's den about an hour after daylight. We found Female 40 still active and away from the den, so Darrell and I proceeded at once into the palmetto thicket. Darrell would earn the distinction of being the most maternal of our crew. Within five minutes he had found the two males in a classic den setting—mostly thick palmetto with staggerbush, saltbush, persimmon sprouts, cabbage palms, wax myrtles, and young oaks scattered nearby. I gathered the squalling kittens into a sack, left Darrell at the den so we could find it again, and took the two kittens to the buggy for measurements and blood samples. (Darrell later shared his vision of being dismembered by Female 40—returning to her kittens only to find a large biologist in their place.) The kittens had open eyes—a milky, deep blue—that apparently could not yet focus. Their ears were unfurled and re-active to sound—both kittens screamed and growled the twenty minutes they were being handled. Their fur was a darker brown than adults, and their black spots were very distinctive. Their tails seemed fairly short relative to body length, and their feet were well armed with white, sharp claws and gray, shiny footpads. Neither kitten had teeth. One kitten received a tattooed Number 1 in his ear, the other a Number 2.

During the workup we could detect Female 40 with our telemetry equipment, but she did not return while we were there. Immediately afterward I carried the two kittens 30 yards through the palmetto thicket to a relieved Darrell. We took a few photos before we left the kittens at their den and re-treated to replace the battery and paper in our remote monitoring equipment. This biologist-in-a-box indicated that Female 40 returned to her kittens about thirty minutes after we left. The chart paper recorded a high level of activity upon her arrival followed by a constant but fainter signal. On our return visit two days later to check battery and paper, we found that Female 40 had relocated her den in another palmetto thicket about 100 yards away.

It seemed strange to leave empty-handed when we had a mandate to re-move kittens for captive breeding. Among GFC administrators in Talla-hassee, however, female kittens had the highest priority, so a male could not be taken unless there was a complement of the opposite sex. Apparently, an oversupply of males in captivity would create a bad public image. Yet males would seem to carry the same amount of genetic material as females. If, in fact, we were taking individuals to help preserve what little genetic variability was left, this litter was crucial. The genes of the father, Male 26, were not represented in captivity with offspring, and he had been a successful and durable resident. His bloodlines would be of value to the budding insurance population.

Nevertheless, we did feel a sense of accomplishment in locating the furry needles in the proverbial palmetto haystack. What impressed me at the den was its spartan condition. No effort had been made by Female 40 to line a nest or in any way provide cushion, insulation, or camouflage for her offspring. Most birds do a much better job of protecting their young. Vertical cover was provided by interlocking palmetto fronds while the prostrate stems of these scrubby palms provided a horizontal background into which the kittens blended. A slight odor of decaying flesh was detectable at the den, but in every other respect it was tidy. A small patch of dark soil was well packed by two weeks of use, and we found a satellite bed site about a yard away. This may have served as a separate, undisturbed zone for Female 40. Neither kitten was capable of walking. Noises unfamiliar to them were met with convulsive hisses that sent them tumbling over with claws extended—a totally ineffective defense, not only against our well-intended intrusion, but against any other potential predators such as bobcats or black bears that might stumble onto the litter. The only real protection these kittens had in their first month of life was the impenetrable palmetto and the wrath of their mother when she was home.

In the following months we would find these den characteristics and panther behavior to be typical. By the end of 1992, I was confident that our studies of Florida panthers had revealed all the general patterns of habitat use, productivity, and social ecology that characterized this endangered population. Our confidence was tempered, however, by recognizing that exceptions can often define the norm. Despite expecting the unexpected in panther biology, we were quite unprepared for Female 48's expansion of these flexible boundaries.

24 February 1992 ■ We first encountered Female 48 when she was four months old and we were still seeking kittens to remove for captive breeding. But because her mother, Number 31, had given up both kittens from her last litter, Female 48 and her sister were safe from a similar threat to their family. Hoping to find the family together, I sent Roy after Number 31 who was located on the northern boundary of the Florida Panther National Wildlife Refuge. At first just the adult female treed. But Roy took his dogs back to a deer kill the family had just fed on, and he picked up the trail of a small kitten that quickly treed in a spindly laurel oak. She was perched precariously (Figure 9.1). When we darted her, she ran out of the tree and disappeared into a lush blanket of ferns, wild coffee, and grapevines. We took off after her but could not keep up, so Roy released Susie, one of his four-legged

Figure 9.1
Female 48 at her first capture.

assistants, to help us. Because Female 48 did not yet wear a collar, it would be hard to find her. And we had to act quickly to make sure she did not have drug complications. Knowing the dog's lethal interest in cats, I was sure we would soon have a mauled kitten on our hands. But Jayde stayed close to Susie who finally barked, then ran *over* the cat, allowing Jayde to scoop up the drugged and helpless kitten. I have no doubt that Female 48 escaped detection when she finally succumbed to anesthesia and collapsed in a fern bed—her motionless body eliciting no immediate response from a dog seeking running quarry. Jayde had surely saved Female 48's life.

After a smooth workup we returned the kitten to the fresh kill site where we suspected Number 31 and her other daughter would soon return. By the next morning, Numbers 31 and 48 were transmitting from the same spot a few miles from the previous capture site. The reunited family, including Female 48's sister who we captured three months later, traveled through Number 31's home range as a unit until the kittens were about thirteen months old. Thereafter we would find the kittens, now independent subadults, increasingly distant from their mother until each was about one female home range width (about 10 miles) away from her birthplace. Number 48's next capture—to replace her kitten collar with one of adult size—happened early in the 1993 capture season.

5 January 1993 ■ Standing atop the 100-foot-tall Miles City fire tower near Interstate 75 and State Road 29, I had just enough elevation to locate Number 48 through early morning fog over the Florida Panther National Wildlife Refuge. With a large group of able-bodied helpers from the U.S. Fish and Wildlife Service, we headed toward the panther on two swamp buggies. During the chase Jayde, Roy, the dogs, and I happened to surround a barred owl perched low in a tangle of oak branches and vines. It twisted and turned its head while watching the dogs tree the cat—then silently took off. The cat found refuge in a lush tropical hammock with huge gumbo limbo, Spanish stopper, wild citrus, red bay and sweet bay trees, wild coffee, myrcene, and satin leaf bushes, swamp fern and strap fern, underlain with treacherous limestone solution holes. Shortly after the dart found its way through a cluster of branches, we caught the drugged subadult female in the net while she was still hot and panting. Number 48's recovery was aided with ice retrieved from the buggy to lower her body temperature. Throughout the workup she fought anesthesia. I only managed to get the full-sized collar on her as she was starting to run off.

Over the next few months, Number 48 used the southeast corner of the Panther Refuge before traveling east across SR 29 and into Bear Island. Here she had her first encounters with adult Males 12 and 26—as well as her first experiences with recreational sport hunting on public land. As far as lands administered by the National Park Service were concerned, the Big Cypress National Preserve was far from typical. Charged with a federal mandate to provide opportunities for hunting and off-road vehicle use on this 700,000-plus acres of wetlands-dominated expanse, the NPS found itself in an uncomfortable partnership with a state agency guided by a very different philosophy of public land stewardship. In essence the National Park Service was engaged in a constant effort to reduce hunting and vehicle activity in the preserve, while at the same time the Game and Fresh Water Fish Commission was obliged to maintain traditional patterns of game harvest and off-road vehicle use. One result of this clash of philosophies was a gradual shift toward conservative management and restrictions on human activities. Another driving force behind increased conservation-oriented regulation was the panther. The National Park Service feared that hunting would not only result in fewer deer and hogs for panthers to eat, but the simple presence of hunters, camps, and swamp buggies would drive panthers away. It was assumed, too, that the churning tires of swamp buggies would alter soils and drainage patterns and damage the overall quality of habitat in Big Cypress. As a result, during the 1980s the use of dogs for hunting deer and hogs was banned, the hunting season was shortened by a hundred days, and all off-road vehicles in Bear Island were required to use a system of designated trails.

Although our studies of white-tailed deer in Bear Island suggested a healthy, productive, and abundant herd in this management unit, it was much more difficult to measure the impact of widespread human activity on panther movements and behavior. Part of the problem was the landscape itself and the problems that dense vegetation pose for monitoring behavior. Easily 99 percent of what we know about panther behavior has been inferred through telemetry, rather than direct observation. What impact would hunters have on panthers? The answer would require a complex experimental design and many more years of research. Large study areas (greater than 30,000 acres) of comparable panther density, vegetation, hydrology, and prey composition would be needed in order to test the hypothesis that deer hunting harms panthers. Although panthers live in an ever-changing continuum of landscape and no single 30,000-acre tract is like another, there are parts of the south Florida landscape that do provide valuable lessons about the way panthers and people tolerate each other. And these lessons could save taxpayers huge sums (and agencies much trouble) by avoiding unnecessary research and potential disturbances to the intended beneficiaries of these actions. All too often wildlife research is funded at great expense when a modest amount of simple interpretation would suffice. Aldo Leopold, in his 1933 text *Game Management,* anchored his classic book with an observation: "Game management is the art of making land produce sustained annual crops of wild game for recreational use." True, the endangered species of today were often the vermin of yesteryear, but that same art is as relevant today as it ever was. Like deer, carnivores such as panthers can be successfully managed to maintain or even increase their numbers. Further, some questions will never be answered by science—and the impact of deer hunting on panthers is probably one of them. Yet wildlife scientists are discouraged from applying the accumulated knowledge of a century of game management experience. Too often they are forced to reinvent an expensive wheel.

The National Park Service trail system for off-road vehicles in Bear Island was intended to concentrate human activity on a well-marked network and leave most of the 38,000-acre area free from random traffic and widespread swamp buggy ruts. Based on our observations, most hunters honored this new restriction. But the concentration of buggy and off-road travel was causing severe rutting, especially when the trails crossed over wetland soils. In some places trails had expanded to 100 yards in width as buggy drivers attempted to avoid axle-grabbing mud that could impede travel and damage equipment. Thus it was hard to assess the trail system's impact on the landscape or wildlife. The debate between the NPS and the GFC continued.

When we least expected it, insight into the troublesome debate over hunting materialized out of the muddy ruts of agency bureaucracy and the ooze of Bear Island's Immokalee fine sand soils. After archery and black-powder seasons had passed, Darrell returned from a telemetry flight on 24 October 1993 and announced that Number 48 was showing signs of den-ning. The last two flights had placed her in the same palmetto thicket in north central Bear Island, and Darrell was rarely wrong in recognizing these natural history events. But a look at the location on the USGS topographic map turned my curiosity to concern. Number 48 had chosen to give birth to her first litter a scant 50 yards away from one of the busiest designated trails in Bear Island. At that moment I realized that Number 48's response to the opening of the general gun hunting season would shed light on the long-standing debate over big game hunting and Florida panthers. In two weeks hunters and vehicles would fill the woods. Many of them would pass within a stone's throw of a litter of three-week-old kittens.

How would she respond? When I informed the National Park Service of the situation, the NPS offered to close the trail in question. I suggested we sit back and see what unfolded. To my surprise, Buck Thackeray of the service agreed not to intervene in any way, thereby allowing us to treat the event as an impromptu experiment. He may have understood the implications as well as I did: abnormal behavior or kitten abandonment by Number 48 would re-inforce the NPS's efforts to move toward a hunting ban. But if Number 48 succeeded in raising her litter, the GFC would feel justified in more vigor-ously supporting hunting. Time would tell.

One piece of the puzzle remained to be put in place. A count of Number 48's kittens and an evaluation of their health would reveal the size of her litter and, subsequently, the number she would be able to raise. We began our efforts to locate these kittens in late October 1993. Darrell and I had already set up our remote monitoring equipment near the den. Mike Dunbar, who had recently replaced Melody after her resignation from the GFC and de-parture for Tanzania, and veterinary assistant Mark Cunningham were down to help us recapture Female 9, so they stayed to help us look for these kittens and collect biological samples.

29 October 1993 ■ We had been trying since 27 October not only to find Number 48's kittens but to recapture Number 9 as well—all to no avail. Number 48 would not leave her den when we were there, and Number 9 went up a huge cypress tree every time we chased her. On this day Number 9 was with Number 51, so we left the aging female alone and con-centrated on Number 48. The pug marks near her den and the lines drawn by the biologist-in-a-box showed that she came and went frequently—we found excellent tracks in the mud of the designated trail a mere 50 yards from her den. We left Mark near the den to monitor Number 48 during the day in case she left. Nothing happened.

Mike and I returned at 4:30 P.M. with the wind picking up. Common grackles were flying and clicking metallically everywhere as they descended upon a heavy acorn crop. Cows mooed nearby. Wild turkeys seemed to be gobbling everywhere. The sky was bright blue with a few puffy cumulus traveling to the northwest—on their way to usher in rain and an early cold front expected to pass through tomorrow. Yesterday I had surprised a flock of wood ducks vocalizing and socializing actively in shallow water at the edge of a laurel oak hammock. Around the perimeter of a large palmetto thicket just north of the den there were bear and panther tracks. And abundant hog tracks, and the overturned soil from their rooting, indicated that prey were easily available to Number 48. As the sun set, though, we made another trip home no wiser for our patience.

30 October 1993 ■ We began our efforts early enough to meet the sunrise but with bleary eyes. I met Mike and Mark at Bear Island Grade and State Road 29 at 5:20 A.M. Maybe by showing up at the crack of dawn we would have a better chance of catching Number 48 away from home. The moon was full and bright as it periodically appeared between clouds that were still blowing to the north. We flushed yellow-crowned night herons from the flooded buggy trail and heard barred owls calling from thick hammocks. Number 48 was not at her den, but she was not that far away either. We would have to wait until there was enough light to see. We swatted mosquitoes as the eastern horizon brightened.

With barely enough light to see, we left the buggy and walked to the edge of Number 48's palmetto thicket. Her radio signal was much stronger now, but she had not quite made it back to her den. Because this was Mike's and Mark's first search for kittens, I spent a few moments explaining the things to look for. The most useful pointer I could provide them was simple: "When the palmettos get so thick you can't go any farther, keep going." And then we headed in. Our search had all the earmarks of previous hunts for dens. But what we found was not what I expected. This was a much more open thicket than the views from the edges suggested. For ten or fifteen minutes we saw no sign of panther presence—no tracks, no trails, no abandoned daytime rest sites. Mike was scouting a tall ridge of palmettos on the north edge of the thicket. At one point Mark and I crossed paths near the slash pine tree I had picked out as a landmark. I shrugged my shoulders. I didn't have a clue where to look next.

At about that moment Mike whistled—our signal—so we hurried to find him. In less than a minute we were huddled behind Mike peering into

Number 48's first den. She had, in fact, chosen the densest part of the thicket. Very little light could penetrate the thick canopy of palmetto fronds. Scattered wax myrtle, staggerbush, gallberry, beautyberry, and vines of wild grape, Virginia creeper, and greenbriar helped conceal a wedge of palmetto roots surrounding less than one square yard of bare, black mineral soil. Looking more like large oblong mushrooms than any kind of animal life were three panther kittens lying motionless on the ground. As Mike reached in to examine the first funguslike fur ball, I wondered where Number 48 was. Actually, I'd been thinking about this for quite a long time. I resisted unfolding my antenna and turning on the receiver, though, for fear of finding out what I didn't want to know. She probably was all too aware of our whereabouts.

These were the youngest panthers we had handled during eight years of kitten captures. Out of three pairs of eyes, only a single eye was open, and this belonged to the male. Applying slight pressure on the eyelids of the kittens indicated they were still well-sealed and not ready to reveal their milky blue orbs. Based on Number 48's behavior patterns of the previous two weeks and the kittens' undeveloped appearance, I placed their age at exactly one week. Surprisingly, one of the females was the largest of the litter at 1 pound 11 ounces. The other female and male were even at 1 pound 6 ounces. None of them was vocal, although they tried to emit a few halfhearted screeches. The male was the feistiest of the three. He squirmed continuously when he was handled. We set up an assembly line: Mike drew a tiny blood sample, then I weighed, measured, and placed a numbered tattoo in one ear of each. Mark was caught in the middle, acting as the transfer point between us. The final step was an injection of Ivermectin, a drug that would help the kittens ward off parasites. I found it amazing that the male kitten's 1.3-centimeter-wide front footpads would quadruple in size in less than two years. In that time his shiny, smooth, tender gray footpads would turn into the cracked, cratered, and coarse black pugs of an adult male. That is, if he survived.

We were through in about a half hour. Our shirts were soaked with a combination of sweat and early morning dew. The humidity was over 90 percent. Stowing equipment into our packs, we made a quick retreat from the palmetto patch. Immediately I unfurled my antenna, turned on the receiver, and was greeted with the loud thumping of a transmitter at very close range. There was no telling exactly how long Number 48 had been in the palmettos with us, but I suspect she had arrived shortly after Mike discovered the kittens.

Florida panthers, like their North and South American relatives, are superbly adapted for killing animals larger than themselves. They know how to use stealth and concealing cover to ambush prey and avoid confrontations with other panthers. Even though we had created a novel situation for Number 48, evolution seemed to have prepared her well for it. The bond with her kittens kept her close to the den, but no aggression was directed our way (though I'm sure she would have won if she tried). Even if she were to win the fight and rescue her offspring, an aggressive encounter with potential kitten predators could leave a mother panther injured and incapable of raising her young. As a result, nature had opted for a strategy of benign neglect in these situations. Black bears too will temporarily abandon helpless cubs at den sites, and the broken-wing act of killdeer and nighthawks is a well-known example in birds of a nonaggressive reaction to a threat. In the event that their young are lost at an early age to predators, at least the mothers will survive to breed again. From the dispassionate eye of Mother Nature, only time will have been lost.

Our biologist-in-a-box confirmed that Number 48 had indeed returned to the den area while we were there working with her offspring. She stayed for about three hours after we left—and then began an erratic pattern of movement that resulted in a steady but much weaker signal than we had received the day before. As the cold front finally passed through the next morning, I flew through low clouds and 22-knot winds to see if there were any measurable effects of our intrusion. Her transmitter pulsed from a palmetto thicket about 200 yards from her original den site on the other side of the designated trail. Because it was so close to the site that Mike had found, I felt confident that she had not abandoned her kittens. Subsequent trips to the biologist-in-a-box confirmed that Number 48 had relocated her den to this new spot across the road. With her new den still within 50 yards of the designated trail, our ad hoc experiment remained intact.

In the following weeks Number 48 and her kittens would be subjected to an array of human disturbances more intense and potentially threatening than any we had imagined. Only her mother, Female 31, had come even close to selecting a den so near a high-human-use area. About four years previously, Female 31 had raised three kittens in a palmetto thicket at the edge of an active cucumber field along Camp Keias Strand. Furthermore, she had selected a spot where tractors, water pumps, migrant workers, and panther biologists maintained a steady buzz of activity at the end of the busy growing season. Though the nature of human activity was quite different, its duration in both cases was the same. About 75 percent of Number 48's denning period overlapped with the general hunting season in Big Cypress National Preserve: 13 November through the end of December 1993. As the most popular management unit in the preserve, Bear Island supported hunters daily, though their numbers varied as the season progressed. On the opening Saturday we encountered many swamp buggies, all-terrain cycles, and people on foot near Number 48's den. She was not there at 8 A.M. Had she already abandoned this site due to the rise in human use? On our next check, two days later, she was at home.

As the weeks went by, Number 48 exhibited the variable pattern of presence and absence that we had seen with all the denning females we had studied. With the rainy season now more than a month behind us, water levels dropped and travel became easier for us. We began encountering Number 48's tracks almost everywhere we drove. Almost every day we would encounter vehicles and hear gunfire as close as 100 yards away from her den to as far away as a mile. The day of 5 December was unusual, however, due to the number of people milling about. While we parked about two-tenths of a mile from the den, we watched ten all-terrain cycles and three swamp buggies pass by within a few minutes of each other. Several of the vehicles were backfiring loudly and carrying passengers engaged in animated conversation. Four cowboys on horseback rode past with their two cattle dogs in train. When we struck up a conversation they assured us the hounds had no interest in panthers. One of the cowboys remarked that he liked the designated trail system because they saw more wildlife and their cattle tended to be scattered less by random buggy traffic. Hunters with rifles crashed through vegetation at the edge of the palmetto thicket. And somehow, in the midst of all this activity, Number 48 returned to nurse her waiting brood.

Two days later we found the tracks of two resident adult males on the trail that passed Number 48's den. These belonged to Male 12 and 26 in a small area where their home ranges overlapped. They seemed to have ignored each other—but more important, we saw no evidence that they lingered near the den. The most serious threat to Number 48 and her kittens occurred while I was in Texas attending an Endangered Cats Recovery Team meeting for the U.S. Fish and Wildlife Service. While I was away discussing ocelot, jaguarundi, and jaguar biology, the National Park Service conducted a prescribed burn in Bear Island—a common land management tool that is used on public and private lands alike in Florida. The purpose of a prescribed burn is to increase the nutritional quality of wildlife and livestock forage, maintain plant communities that depend on periodic fires, and reduce the dead vegetation fuel loads that could support a dangerous wildfire. Darrell told me over the phone that he had seen smoke rising all around the den during his telemetry flight on 10 December. The next day I verified for myself that the fire had stretched for about a mile, starting at Bear Island Grade and ending no more than 50 yards from the den. Why this highly flammable patch of palmettos did not burn is anyone's guess. But just as we had seen with the previous weeks of disturbance, this one seemed to have no effect on Number 48. On 11 December she was at home, apparently no worse off.

On the way home from Bear Island I stopped by Darrell's house to see if he could give me more details. After all, even when they are green the leaves of saw palmetto are extremely flammable. In a small prescribed fire several years ago at my home in east Naples, we watched as palmetto-fueled flames leaped upwards, setting branches of living pine trees on fire 20 feet above the ground, and searing the hair from my wife's eyebrows as we scrambled to keep the flames in check. At least we could have retreated. The kittens, however, would not have left the large palmetto thicket unless they were carried out by their mother. Although professional fire managers have elevated prescribed burning to a high art, even the most experienced of them acknowledge that unpredictability is an important ingredient in any "controlled burn." A sudden shift in wind strength, wind direction, or moisture conditions can doom a burn to a fizzle or whip it into a roaring conflagration. Fires in green palmettos do not creep—they roar. Amazingly, through blind luck or divine intervention, the den and its precious contents escaped conversion to ashes.

Panther 48 tended her den and kittens through 21 December. Her departure was preceded by at least two visits to the Hendry County–Collier County line about a mile away. We recorded her last visit to the den as she gathered her litter and led them to their first carcass at the edge of a complex of farm fields.

*   *   *

It is tempting to view Number 48's tenacity as an amazing feat of animal adaptability. Yet there are many other examples of wildlife and humans sharing space harmoniously. And among mammals, none have filled the wide variety of ecological opportunities in the Western Hemisphere better than *Felis concolor*. Only where humans totally dominate the landscape have panthers been eliminated from their former range. They seem quite tolerant of moderate levels of cattle ranching and the presence of people—from Florida to Texas, from northern California to Mexico, from Colombia to Patagonia.

One key to this peaceful coexistence is *regularity* in the human activities. Among birds, the usually elusive American bittern can be seen close up and undisturbed from the boardwalk in Everglades National Park's Anhinga Trail, where people move predictably along an unchanging pathway. Many avian species from wood ducks to roseate spoonbills can be seen feeding contentedly in the drainage canals of Naples while cars and trucks rumble past. Among mammals, elk in Gardner, Montana, adjust to day-to-day town life and African lions in the Serengeti ignore the intrusive Land Rovers and gawking tourists. Peregrine falcons nest on buildings in Baltimore, black bears den under houses in Pennsylvania, burrowing owls live in grassy midways of airports, purple martins nest in the moving booms of giant oil derricks, alligators sun on the banks of busy golf course water hazards—the examples are endless.

Is it so surprising, then, that a female Florida panther raised her first litter a mere stone's throw from a regularly used swamp buggy trail? In light of examples of greater adaptability by species of lesser brain power, Number 48's accomplishment seems less spectacular. But more than luck played a role in her success. Human activity in Bear Island occurred only during the day, and most hunters confined their buggy travel to the new system of designated trails. No longer did hunters crash through palmettos in high-wheeled buggies to flush out their targets. Number 48 had the nights to herself and managed to adjust to the daytime activity patterns of people. If any debate was warranted, it should have been over the *kinds* of hunting allowed where panthers live—not over whether managed hunting is acceptable in a national preserve that was legislatively mandated to allow and perpetuate hunting. A fifty-year hunting moratorium had not stopped panthers from going extinct in Everglades National Park. But uncontrolled human activities in the southern Golden Gate Estates have probably done much to prevent panthers from recolonizing these 40,000 acres. Somewhere between these two extremes—total control versus no control—is a level of human activity that is compatible with panthers. Number 48 may have shown us this level.

DISTURBING 10
REVELATIONS

Number 48 may have demonstrated that hunters were not insufferable pests. But her tolerance of people with guns and swamp buggies caused me to wonder if we had witnessed other situations suggesting a peaceful coexistence between humans and panthers. Perhaps by examining the impacts of a variety of disturbances a pattern would emerge. Surely the species had weathered enough changes, even before the advent of humans, to survive a few adjustments to its landscape. One of the first of these disturbances was a purely natural one known as Hurricane Andrew.

Just over a year before Number 48's experiences with a Bear Island hunting season, we had the opportunity to evaluate one of nature's most awesome displays of power and violence on the planet—a hurricane—and to better appreciate a big storm's ability to decimate panthers. With the tropical storm season in full swing, and kitten removal efforts continuing, the timing was ideal for witnessing a natural disturbance that has long been touted as one of the many threats to panther survival, even though no such disasters have been the ultimate cause of extinction in a wide-ranging carnivore. Although south Florida is always a prime target for summer hurricanes, none had hit panther habitat since the inception of research in 1981. No one knew what a category five hurricane would do to south Florida's interior and its wildlife. Would the devastation that an extremely powerful storm could swirl onto a landscape susceptible to floods and wind be enough to push the panther below some unknown critical threshold?

19 August 1992 ▪ Because the litter of Female 23 had been targeted for captivity, I was responsible for coordinating kitten removals with the National Park Service even though we were unfamiliar with the movements of these animals and were generally unwelcome in the Big Cypress National

Preserve. Ever since the beginnings of panther research on NPS lands, the monitoring of panthers had become an increasingly complicated affair. Everglades National Park had its own research team, while the Big Cypress National Preserve fielded yet another crew for panthers captured there—even though study animals from both areas crossed back and forth from one federally managed tract to another. The one common thread in these duplicate efforts was the need for the GFC field crew to oversee all panther captures. In addition to Deb Jansen, the preserve's wildlife biologist who monitored the handful of panthers living in this sub-par habitat, four other National Park Service employees accompanied us on a long swamp buggy trip from Oasis Ranger Station to Raccoon Point. My basic crew consisted of Darrell, veterinarian Carolyn Glass, and her assistant Pauline Nol. Bill Greer, a GFC photojournalist, and David Villano, a Miami-based freelance writer, were also along to document our controversial kitten removals.

Our first attempt at kitten finding was stymied by Female 23's presence at her den until late afternoon. A nighttime search would likely have been unsuccessful and disruptive, and could have prompted Female 23 to move her den, so we rescheduled for the next day. This did allow us, however, to get a better feel for the den's location by closely monitoring Female 23's radio signal and selecting landmarks to guide us when she finally did leave during the day.

20 August 1992 ▪ The National Park Service contingent was down to a crew of one today, perhaps due to lessened expectations, and Jayde took Darrell's place on my team. This group was a bit more reasonable in size, but still more than necessary. Our most efficient efforts to locate dens came when just two of us looked for the kittens. As we approached Female 23's den at about 5 P.M. we noticed that her signal was much fainter and more to the southeast than yesterday's location. Her den was unattended and the weather was threatening. The rumble of thunder and occasional lightning flashes reminded us that south Florida was still in the dank clutches of summer. The pattern of afternoon thunderstorms would not be broken until late September.

Wanting to make the most of remaining daylight, beat Female 23's return, and still avoid the arrival of driving rain, we began our search at the tip of a long peninsula of hardwood hammock surrounded by cypress swamp, wet prairie, and open pine woods. The thunderstorms continued to build around us and the mosquitoes were worse than the day before, as Deb, Jayde, Carolyn, and I searched for the den. Light was subdued from the overcast and it was very humid—I was drenched with sweat in minutes. The hammock turned out to be much larger than our views had indicated yesterday.

We found nothing at the base of the oak where we expected to find the female—maybe one mashed-down spot on dry leaves—so we expanded our search deeper into the hammock but still found little sign of large animal activity. At one point I felt the search was useless. The area was so much bigger than I had first thought. But after about fifteen minutes of checking every nook and cranny, Jayde announced that he had found a matted area at the edge of a large fern bed. I worked my way toward him.

As I passed through the calf-high ferns there appeared to be a sharp break in the dense foliage. And, indeed, the ferns were freshly flattened. A glance to the left revealed a spotted fur ball tucked between two downed palms on matted fern leaves. Quietly I signaled everyone to the verdant alcove. The two kittens were oblivious to our presence until Jayde touched one causing it to jump and snort slightly. I estimated their ages at eight to ten days: they seemed much less developed than the first litter we had handled, Female 40's fourteen-day-old males, and they did not vocalize (except when Carolyn used a needle to draw blood). Within five minutes there was a calm buzz of activity as both were weighed, measured, and photographed. The female had a tick on the hairless skin above an eye, and the tissue around both eyes appeared slightly puffy. The male had a hairless ring about 3 inches from the base of the tail where it had received some sort of injury or irritation. They did not appear as thrifty as the others we had handled on lands further north. White blisterlike circles were visible on their otherwise shiny footpads, and the kittens had a distinct but not overpowering odor to them—mostly like the smell of rotting flesh. This odor was transferred to our hands as we held them, and I wondered if it came from the male with the irritated tail.

Rain sprinkled lightly as we left the den. Jayde and I each carried a kitten wrapped in a small towel (Figure 10.1). We could not help feeling some compassion for Female 23 who would discover the loss within the next day. Perhaps the kittens knew too little to notice the change. In any event, by midnight they would both be introduced to their new home at the Jacksonville Zoo, 300 miles to the north.

On my way home that evening a radio weather report indicated that a westward-tracking Atlantic tropical depression had been named tropical storm Andrew. In just over a day it would grow into a hurricane, Homestead, Florida, would be devastated, and extensive damage would be wrought on south Florida's mangrove and native slash-pine forests. And because most people mistakenly equate panthers with the Everglades, it was assumed that panthers had taken a beating from the force of the storm. A news item from the *Miami Herald,* showing a photo of one of the kittens we had removed, reported that the young panthers had been "rescued" because of the threat of Andrew.

Figure 10.1
Jayde Roof cradles one of Panther 23's kittens in the eastern Big Cypress National Preserve.

In fact, none of the radio-collared panthers altered their movements after the storm. My guess is that they viewed the events of 24 August 1992 as nothing more than a severe thunderstorm. Even if the hurricane had tracked 30 miles further north and through the heart of panther range, I doubt we would have noted any changes in panther behavior. I had always considered the notion of panther-killing hurricanes a little melodramatic anyway—dense palmetto thickets are probably very safe places to ride out such a calamity. While hurricanes such as Hugo in 1988 have devastated local populations of cavity-nesting birds, including red-cockaded woodpeckers in Francis Marion National Forest, South Carolina, large solitary mammals are much less vulnerable. Without the walls, ceilings, glass, or furniture of civilization to become lethal weapons, the dense palmetto thickets selected by panthers as dens and daytime rest sites are probably fairly immune from high wind and driving rain. Not only are these sites extremely well drained, but there are no heavy branches and such to be thrown about. One of our radio-collared black bears, a 300-pound male, experienced the eye of Hurricane Andrew from his spot in dense mangroves on Everglades National Park's Pavilion Key. Despite much wind damage to the mangroves, this bear's behavior suggested that he was oblivious to the storm. I suspect that panthers, including Female 23 and her kittens, would have survived a direct hit.

26 August 1992 ■ On a telemetry flight two days after the storm I had a chance to view the panther landscape firsthand from 500 to 1000 feet above the ground. During the flight over the heart of occupied panther range, I noticed nothing that might have indicated a significant response to Hurricane Andrew. The most noticeable alteration was the royal palm fronds in the Fakahatchee Strand pointing to the west. Several clumps of cabbage palms looked frayed from having their fronds whipped about. Oaks, cypress, and maples had lost leaves throughout south Florida, and stout branches were stripped like toothpicks from a number of trees. Nonetheless, except for large accumulations of understory debris, the forests of the panther maintained their thick character.

Phone inquiries poured in that day from reporters throughout Florida and from as far away as Washington and Boston—an indication of how widely the public was aware of the Florida panther. I responded to their questions by telling them that I was glad, but not surprised, that all of our panthers had been accounted for and were doing well.

*   *   *

While Hurricane Andrew was a dramatic test of wildlife resiliency, it was very different from the kinds of human disturbances that are equated with the development of south Florida. These projects slowly change the landscape by chipping away at habitat or altering the way a piece of land is managed. Florida's media provided in-depth coverage of a controversy that had begun two years before I arrived in south Florida—one that led to a purely unnatural disturbance in the heart of panther range. During 1983 and 1984 a flurry of newspaper stories, many by *St. Petersburg Times* reporter Pete Gallagher, covered the plans of Ford Motor Company to build a vehicle evaluation center in central Collier County—just west of the current Florida Panther National Wildlife Refuge and just east of the Golden Gate Estates. Looking back on the articles that chronicled this strange choice for the site of a high-tech facility, I am amazed that it was ever approved by the commissioners of Collier County and that state and federal agencies issued permits for its construction. When completed in 1984, the compound would support a 2.1-mile high-speed straightaway, a 1.1-mile handling course, and several other road surfaces designed for suspension tests, sound measurements, and other experimental activities (Figure 10.2). To eliminate the chance of a dangerous collision between a large animal and a test vehicle, it would have a fence to exclude panthers and other large wildlife species from entering the 530-acre facility.

Before construction began, the land was predominantly composed of cypress swamp, pine flatwoods, and hardwood hammock—ideal panther habitat. In fact, two locations of radio-collared panthers had been documented on the property that was, and still is, owned by Collier Enterprises. From a natural resources perspective, the land was in a strategic location: a buffer between the rapidly developing northern Golden Gate Estates and wilder lands to the east. From a vehicle development perspective, the facility was far from heavy urban traffic and provided security for the development of prototypes. Even after construction, more than 460 acres of forest remained, some of which existed as spreading live oaks and other trees that responded to increased edges and sunlight by draping their branches over the perimeter fence. As a result, the facility maintained most of its natural amenities.

Collier County has a long history of questionable environmental decisions. In fact, its approval for the Ford project was given despite universal objections by the environmental community and after the Board of Commissioners had initially rejected it. Apparently the allure of nonindustrial development, the lack of coordinated opposition, and development-friendly local politics drove the project to its completion. In the end, nearly a square mile of forest adjacent to Alligator Alley was fenced with chain link and topped with three strands of barbed wire outrigger. This, it was argued, would be a minimal loss of habitat. And the fence would prevent wayward panthers from blundering into the path of a red-lined Thunderbird. In reviewing the files on the evolution of the "test track" saga, it is clear that the politicking that preceded the granting of permits could supply enough material for someone's dissertation on Collier County development practices during the 1980s. More important, the appearance of this facility in occupied panther range provided yet another way of measuring the Florida panther's ability to tolerate change.

For the most part, Ford's big move to Florida was reported thoroughly by the news media. They missed out on the best and most ironic part, however, after all the editorial hyperbole and furor had become fish wrap. The epilogue to this story was even stranger than the politics that went into the

Figure 10.2
The Ford Vehicle Evaluation Center, I-75 (*foreground*), and the Florida Panther National
Wildlife Refuge to the east. The "wildlife-proof" fence surrounding the facility and hun-
dreds of acres of native forest did little to keep panthers out.

track's approval. Eight feet of chain-link fence, three strands of barbed
wire, asphalt, concrete, and Mustangs notwithstanding, several panthers—
including at minimum Numbers 10, 25, and 32—chose to visit the "wildlife-
proof" compound about two dozen times over the next several years. In
addition, we found that radio-collared black bears were entering and using
the facility even more frequently than panthers. Through ten years of operation, no bears or panthers are known to have been hit. Not even a close call
has been reported.

As unlikely as it sounds, the facility actually offered some benefits to
wildlife. Although the fence was a dismal failure in keeping animals out, it
was quite effective in limiting access by people—an artifact of Ford's desire

to maintain secrecy around new car designs. So far as I know, Ford employees never ventured into the heavily vegetated pine flatwoods, cypress swamps, and hardwood hammocks within the compound. On the few occasions that we visited the facility to check on a collared panther, we found abundant sign of wildlife, ranging from panther kills to acorn-filled black bear droppings. Further, because the activities related to vehicle evaluation occurred most often (and predictably) during the day, and the facility all but shut down during the summer, it was not hard to understand why large secretive mammals might be drawn to this strange sanctuary. Just as Female 48 had adjusted to human activity in Bear Island to raise her first litter, panthers and other wildlife had accepted the human intrusion of the test track and continued to use the excellent habitat contained within the fence.

In retrospect it is probably safe to say that a better construction site could have been chosen for the Ford Evaluation Center. On the positive side, the future Panther Refuge gained a buffer with the protection of nearly 200 acres of intervening swampland known as Lucky Lake Strand—a requirement placed on the owners of the property, Collier Enterprises, as part of the mitigation for the development. More important, it was useful in providing one more testament, not only to the adaptability of panthers and other wildlife, but also to the fallibility of humans in predicting the future. Surely no one could have imagined panthers *and* bears slinking and lumbering over the fence and through the steamy hammocks within the fenced compound while the next generation of passenger vehicles hurtled through the middle of this forest fragment. But use it they did—and in numbers that were indistinguishable from pretrack conditions.

* * *

As researchers, our most important tools in the field were those that allowed us to capture panthers. Thus Roy's hounds, in addition to being very personable dogs, were tops on our list of equipment. How ironic, then, that hunting with dogs has been blamed, in part, for the panther's dismal survival status today. Modern panther hunts are not so different from the killing expeditions of a previous age—it's just their outcomes that have changed.

Commission biologist Chris Belden, author Jim Bob Tinsley, and others have suggested that unmitigated persecution in the nineteenth and early twentieth centuries led to the Florida panther's demise. This is a fairly safe, if not necessarily accurate, statement. After all, the alleged perpetrators of this persecution are long gone, and it is too late to measure the depredation. While many panthers certainly took their last breath in the top of a tree with a circle of baying hounds on the ground, these panthers became accessible primarily as the result of forest fragmentation and expanding road networks.

In other words, apparent panther eradication in north and central Florida came only after the human invasion and settlement of the state's interior. Many of the difficulties experienced by the Texas cougars—released into north Florida in 1988 to test the potential to reestablish a panther population outside south Florida—were directly related to the ease with which people could move about in the woods. Although these experimental immigrants came from a dry and mountainous environment, their problems in north Florida had little to do with climate or geography. They were made vulnerable by a landscape fragmented by urbanization, split by agriculture, and interlaced by a network of roads through private timberlands and national forests.

Much of south Florida has been spared this degree of intrusion. Much of it still presents formidable barriers to human access and travel. Our experience with Roy and his dogs trained specifically for hunting cats suggests that where forest cover is extensive, panthers may be easy to verify but can be extremely difficult to capture. Even where their sign is abundant, great effort precedes the treeing of a single panther. Had Florida's interior not developed in leaps and bounds, had it retained expanses of wilderness comparable in size to south Florida, panthers today would be more widely and securely distributed. And they probably would be able to tolerate a limited hunting season. I would even hazard to suggest that today's remnant population could stand a low level of legal hunting that targeted nonessential individuals—that is, nonresident males. But no matter how biologically defensible this concept may be, today it is morally and socially unacceptable.

Although the killing of a treed, helpless animal is a cowardly act, this hasn't stopped everyone. The few documented cases of modern-day panther poaching resulted from chance encounters between unlucky panthers and humans who, although armed to kill, were not seeking to shoot panthers. Such acts never appear to be premeditated. Of the forty-eight panther deaths recorded between 1978 and 1994, only five were the result of shootings and none occurred after 1985. The only shooting that came without question from the heart of panther country was the celebrated case of Seminole Indian Tribal Chairman James Billie's 1983 trophy. Billie would eventually be exonerated of a crime because a panel of experts could not convince a judge and jury that we knew exactly what a Florida panther really was. Three of the other poached panthers were shot in the relatively poor habitat of southeastern Florida. Perhaps this more accessible grassland-dominated landscape made the few panthers that roamed here more vulnerable than their relatives living in the heavily forested and impenetrable lands further west.

During the course of fieldwork we met many south Florida outdoorsmen, most of whom had stories to tell of their own panther encounters.

Some of the most compelling came from the few cat hunters who still kept and ran their dogs after bobcats. Telling us that a cat hound in southwest Florida chased only bobcats, though, was like Roy saying his hounds chased only panthers. No doubt these few "bobcat" hunters had treed many a panther over the years. And one of these errant chases may have led to the nonfatal shooting of Female 9 in the southern Golden Gate Estates sometime during 1986.

Dick Highmark,* a Collier County vegetable farmer and cat hound hobbyist, was probably second only to the McBrides in the number of panthers treed in south Florida. Highmark had taken an interest in panthers during the early 1980s and enjoyed finding their sign on a large tract of forested land he leased from a local corporation. The property, located adjacent to the Florida Panther National Wildlife Refuge, was excellent deer and bear habitat, and he was quick to take credit for "raising" the panthers that frequented it. My first encounter with Highmark came on 27 January 1987 after we determined that young Male 10 had died just west of State Road 29. Highmark saved us a long hike through very wet mixed swamp by shuttling us to and from the panther, which lay beneath a young cabbage palm along one of the property's many winding trails. The sign of Number 12 and an uncollared female were fresh and abundant near Male 10's dismembered carcass. There was little to salvage. The cause of death was obvious—severe trauma inflicted by Number 12, but we still stuffed his bones and sloughing hide into a black trash bag to save for posterity.

For several years thereafter Highmark took great interest not only in panthers but in how we did our work. He was on hand in 1988 when we unexpectedly captured Number 25 on his lease. He made the curious remark that this young adult was not the killer of a panther whose carcass he had found some months before (but had neglected to tell us about). We would never learn how he "knew" that Number 25 was not the culprit. He did, however, produce photos of a dead and decomposing uncollared panther that could have been taken anywhere. Highmark was so enamored with our work and technology that he outfitted his own hounds with transmitters tuned to the same frequency band we used on panthers and Roy's dogs. On more than one occasion we were confused while aerial tracking when two transmitters emitted signals on identical wavelengths—only their pulse tones allowed us to tell them apart. Walt picked up one of these pesky collars at the edge of State Road 29 near a dog's carcass, presumably one of Highmark's. But more disturbing than the nuisance was the fact that Highmark

*Not his real name.

was capable of tracking our radio-collared panthers. This kind of infiltration into research projects had led to the deaths of many radio-collared black bears in the Appalachian mountains—tracked to their dens and shot for "sport" or for the harvest and sale of bear organs and other parts. Would this happen to panthers? By the end of 1989 we had heard several stories about Highmark showing off photos of a treed collared female and her single kitten. The timing of the rumor suggested that it was Female 19, who was raising a kitten at the time.

6 September 1990 ■ At around midday I received a call from Jack Queen, a local contractor, who related that he and Dick Highmark and his hounds had chased a panther up a pine tree near Queen's home on Lake Trafford. Would we be interested in catching it and collaring it? I told him I would be there in an hour. I wondered how they had gotten tangled up in another panther situation.

On my way to Immokalee I alerted our regional office and requested the assistance of a wildlife officer to deal with a possible endangered species violation. By 2:30 P.M. we were gathered at Queen's concrete driveway, where he explained it was all a mistake. It was a *bobcat* that had killed an orphan deer fawn he had been raising. The fawn was kept in a large 8-foot-tall hog-wire enclosure. We were told that the bobcat killed the deer, dragged it over the top of the fence, and escaped into the woods. He then called Highmark to seek his advice. Highmark suggested they put out his dogs and "teach the bobcat a lesson." So they released Highmark's hounds and almost immediately had not a bobcat but a panther up a tree.

After this introduction Queen escorted us to the deer pen and the scene of the encounter. The sand outside the enclosure was peppered with panther tracks. We did not see a single bobcat track. He then led us about 100 yards into the woods behind his house where Highmark awaited with napping and disinterested hounds and his treed trophy. The huge slash pine rose over the short laurel oaks and saw palmettos. Resting comfortably on a large horizontal limb was a young male panther. Highmark asked if Melody was on her way, hoping that we would take advantage of his good Samaritanism by radio-collaring this animal. I would have liked nothing better than to collar the cat. The Corkscrew Swamp area sorely needed more telemetry data to reveal the true importance of this part of the panther's range. But Roy was not in the state, Melody was five hours away, and there was no way anyone in Tallahassee would let me carry out the capture without them. More important, our acceptance of this "gift" would encourage future pursuits of panthers. After many photos we all left the tree to allow the panther to get on about his business of looking for a vacant home range. The wildlife officers on the scene asked Highmark not to do it again.

11 March 1991 ▪ The last entry to the Highmark files came during the 1991 capture season as we began to remove kittens for captive breeding. The morning started out like most other kitten capture days. Darrell and I waited in my truck at the north boundary of the Florida Panther National Wildlife Refuge for word from Jayde about female panther locations. We watched Jayde circle over Female 31—near her previous day's location—but he saw her walking quickly along the edge of a thick swamp. Apparently she was not with her kittens. As he banked, Jayde spotted a pack of dogs and a pickup truck parked near Highmark's boundary with the Panther Refuge. Several dogs were, in fact, on the refuge and Male 12 was also very close. Jayde tuned in to a few of their frequencies and passed them along to us so we could monitor their progress from the ground. I radioed our dispatcher in West Palm Beach who then notified the refuge manager, Todd Logan, in Naples. He and his assistant, Ken Edwards, headed out immediately to meet us.

Walt and Roy found two more of Highmark's dogs at the edge of a large vegetable farm. Meanwhile Darrell and I listened to barking dogs and a confusion of radio signals emanating from the Panther Refuge. When Todd arrived we decided to head in on foot, wading across a waist-deep drainage ditch toward barking dogs and conflicting radio signals while Ken hurried in on a swamp buggy to find Highmark. The pop-ash ponds we forced ourselves through contained very cold water. (The temperature last night had dropped to 31°F.) The terrain then changed to open cypress, their knees hidden by thick, burr-laden caesar weed, an exotic plant that is famous for its Velcro-like seed capsules which cling to any fabric or hair-covered surface. After a mile or so we intercepted Ken and Highmark not far from barking dogs and a very strong signal from Male 12. It sounded as though the dogs had treed Panther 12—but a moment of silence followed by renewed barking and a fading radio signal indicated the chase was back on. Eventually we caught up again, but this time Male 12 was waiting 25 feet up in a spreading laurel oak. He appeared to be in excellent condition as he looked down on us. Nervously Highmark rounded up his dogs, told them to shut up, and assured us he was "double sorry" for the mistake.

Because this trespass and "harassment" of an endangered species occurred on the federally owned Florida Panther National Wildlife Refuge, we left the prosecution of the case to the district attorney's office in Fort Myers. Aside from the depositions made by Jayde, Darrell, Todd, Ken, and me, though, I am unaware of any other official actions with respect to Highmark's "bobcat" hunt. We were all frustrated that he was using his dogs in this heavily traveled part of panther range. Even if he never felled one with lead, there must have been cumulative impacts of his hunts on panthers. Given the number of times we had caught him red-handed, many more panther chases must have gone unnoticed. Maybe this was a regular weekend activity when the weather was nice. Maybe he rationalized that he was doing nothing different than what we did during our routine captures. Besides, there was nothing illegal about hunting bobcats with dogs. And unless we could prove malicious intent to disturb panthers, this would always be a convenient excuse if caught.

Clearly, Dick Highmark helped to demonstrate how handcuffed wildlife law enforcement can be. But more compelling was evidence of the panther's adaptability and tolerance of yet another form of human disturbance. So far as we knew, none of Highmark's chases had resulted in a dead panther. And if he was regularly hunting panthers in a nonlethal manner, telemetry data from his potential quarries provided us no clue. No home range had been abandoned in this area. Nor did any of our collared study animals exhibit unusual movements. The area produced more than its fair share of panther kittens, and the sign of uncollared cats showed up from time to time. In an area like this with abundant food and cover, panthers may be more capable of withstanding the occasional inconvenience of being treed than in poorer habitat such as the southern Golden Gate Estates and the Fakahatchee Strand. In these areas, where food is limited or escape cover is sparse, the energy a panther spends to evade dogs and climb a tree may be the difference between raising a litter and aborting a pregnancy—or it may be the drain that turns a low-prey-density area into one that is totally uninhabitable by panthers. All I know is that in the one place where we had some evidence of "unofficial," nonlethal panther hunts, there did not appear to be a measurable cause and effect.

*   *   *

In sum, then, our observations of panther reaction to disturbance indicated that members of this endangered subspecies form quite a tolerant lot. Florida panthers, like their cougar kin, make excellent wilderness creatures, but they do not require wilderness for their survival. Natural disasters seem more likely to threaten human lives and property than panthers and their habitat. Moreover, a broad range of human activity is endured by panthers so long as food and cover are adequate. Researcher Paul Beier found that California cougars tended to stay away from well-lit urban areas but did not have an aversion to using concrete structures such as culverts and underpasses for traveling. A rise in attacks on humans in the western states has been attributed to an inherent adaptability of the species that has been challenged by the expansion of human residential areas into historical cougar range. The trend of building larger houses on larger, wooded, naturalistic lots means that good deer habitat and good stalking cover for cougars can be maintained—even in the shadow of $500,000 homes. In south Florida, panthers have shown a resiliency to humans that rivals their western relatives.

NUMBER 44 AND THE PANTHER GAUNTLET

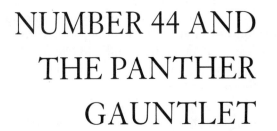

Young male panthers were the ones most likely to show up in someone's backyard or in some other unexpected place. A glimpse into the life of a dispersing male is the best way to illustrate the influence of the landscape and the control that panthers can exert on other panthers. Dispersal of young is a characteristic of all biological populations. But in panthers, as in other large mammalian carnivores, the value of a youthful individual is determined by the rate at which resident adults let them into the fold of panther society. Panthers did not simply blow across the landscape like willow seeds—landing at random and taking root where they came to rest. As we had seen time and again, male panthers had to be extraordinarily lucky to find that bare patch of soil—a place not already defended by an adult.

Number 44 epitomized these difficulties when he attempted to find a place in the tightly woven fabric of panther social structure—the result of long-standing resident males who refused to yield to younger competitors. Compared to the regular travels of resident adults, dispersing males exhibited movements more like the chromium ball in a 1960s vintage pinball machine. They seemed even less predictable than a windborne dandelion seed. In western populations of mountain lions that are subjected to legal hunting, home range vacancies are continually created, providing many opportunities for locally produced offspring. Some researchers suspect that high turnover may also reduce the likelihood of aggressive encounters between males. This appeared to be the case in Maurice Hornocker's landmark study of mountain lions in the Idaho Primitive Area during the late 1960s. But even if no vacancies are available to dispersing males, some parts of the west are still so vast and remote that a juvenile can travel hundreds of miles in any direction to search for one. Such is not the case for panthers in south Florida.

Number 44 was born on Halloween 1990 in the heart of south Florida panther range. At independence, he would have his choice of any dispersal direction. His mother, Female 40, chose a dense palmetto thicket on a privately owned ranch in central Collier County to give birth to her first litter. Her encounter with Male 26 some three months before had resulted in two kittens that we subsequently captured in the spring of 1991. Number 44 was collared and released as a 33-pound, six-month-old kitten on 30 April. His capture was uneventful, and he quickly reunited with his mother. This was the year we began the experiment to establish a captive population while gambling that no long-term impacts would be felt by the wild, free-ranging population. Number 44's litter mate would become part of this captive safety net. One week after capturing Number 44, we captured his sister—the last of six kittens destined that year for White Oak Plantation and other sites in captivity.

Female 40 searched for several days after the loss of her only female kitten (Number 206) before resuming the regular movements of a busy Florida panther mother. She and Number 44 stayed together for another ten months before the family bond dissolved and the youngster began his new life as an independent male with no social standing and little experience in panther etiquette. His initial exploratory movements centered on the familiar territory of his kittenhood—steamy pine flatwoods and isolated, grassy ponds on private lands just north of Interstate 75. Gradually, over a month, his movements expanded and then took on a distinctly westward trend. His travels to Bear Island and the Panther Refuge brought him into potential conflict with resident Males 26 and 12 and led him to parts of the south Florida landscape unfamiliar to him. He traversed State Road 29 and utilized one of the new wildlife underpasses constructed for panthers to cross beneath the busy interstate. By March 1992 he had left permanently occupied panther range and entered the infrequently used southern Golden Gate Estates, just west of the Fakahatchee Strand. Number 44 had successfully negotiated the gauntlet—the most densely occupied habitat in panther range. Now, at about eighteen months of age and with a long dispersal behind him, we assumed he would settle into the temporary holding pattern typical of the other young males we had studied.

The area chosen by Number 44 was a zone of embattlement among private property rights advocates, natural resource agencies, and the Collier County Board of Commissioners. The state of Florida's desire to purchase and protect the remaining intact portions of the western Big Cypress Swamp was being met by strong opposition from a group of Collier County administrators and others recognizing this area's real estate potential. The southern portion of Golden Gate Estates represented the best opportunity to manage native habitat in such a way that panther numbers would actually increase. Up until now all land acquisitions, whether by state, federal, or private interests, had simply secured land already occupied by panthers. When viewed from the air, the southern Golden Gate Estates is a gridiron of crisscrossing roads and south-draining canals etched into a forested landscape. Although few permanent homesites existed here, other human activities such as dog hunting, target shooting, fishing, and garbage dumping went unregulated. In addition, the paved roads with no telephone or power lines were reputed to provide quick landing and takeoff for planes with illicit cargo. Suitable panther habitat existed throughout the southern Golden Gate Estates, but it was underutilized by panthers. No doubt this area would become permanently occupied by resident panthers if human uses were less intense.

Number 44 continued westward into Belle Meade, a 41,000-acre tract just east of the city of Naples, and left the Golden Gate Estates behind. Several other male panthers spent portions of their dispersal wanderings in this mostly forested tract, but in every case they had returned to the east—to less fragmented habitat, more remote forest cover, fewer people, and more panthers. Apparently Number 44 had not yet ceased his explorations.

22 April 1992 ■ I had the radiotelemetry flight on this clear spring day and was in the air with Pete Baranoff, our Cessna pilot, by 6:45 A.M. Turning on the receiver as we crossed County Road 951, I expected the steady sixty-beat-per-minute pulse of Number 44's radio collar to fill my headphones. But we also knew that young males were quite unpredictable in their movements, so I wasn't surprised when his signal did not turn up where expected. As I checked off bears and panthers on my list and continued east toward the Miami side of the Big Cypress Swamp, I began speculating to myself about Number 44's fate. I could think of no panther that had traveled 50 miles in two days. And rarely had a transmitter failed without some hint of a malfunction. With all collared animals but Number 44 accounted for, we found ourselves searching increasingly peripheral areas. But the number of detours we could take to locate this needle-in-a-haystack was limited by the plane's fuel capacity. Pete finally tapped my shoulder, pointed at his watch, then banked the Cessna back to the west. Could Number 44 have become the only radio-collared panther illegally killed, his collar removed and destroyed by poachers? (This was a fairly common fate for south Florida black bears.) As we returned to the Naples airport this possibility seemed increasingly likely.

As I stowed the maps to prepare for landing, I continued to monitor the wayward panther for the odd chance that I had missed him on the way out. With static filling my ears and the airport in sight, it was time to begin our

final approach. I reached to disconnect the coaxial cables that linked my electronics to the strut-mounted antennas. A faint beep, repeated then lost, par-
alyzed my hand. With suburban residences, an Oldsmobile dealership, a
shopping center, and a busy Naples street beneath us, surely someone was
playing a joke on me. Walt, Jayde, and Darrell all had a well-developed sense
of humor—and besides, it had been almost a year since they planted that bear
collar in the woods next to my house while I was away on one of my flights.
Why not a panther collar hidden within a bush on a development in the
fastest-growing part of the United States? The signal increased in intensity
and regularity as we approached the runway. Pete must have questioned my
sanity as I sat up straight and pointed for him to circle. When he shot me a
look of disbelief, I shrugged my shoulders and said "I don't believe it either."
The signal was emanating from a narrow strip of slash pine sandwiched between two recently constructed housing developments—not a bad place to
hide the collar of a panther hit by a car the day before or in some other way
retrieved without my knowledge. At least it wouldn't be hard to find.

I thought Darrell put on a pretty good act when he feigned surprise at the
location of Number 44's signal. Jayde and Walt had probably left the office to
avoid bursting out laughing at my gullibility. With Darrell in tow, I drove to
Embassy Woods ("Homes from the $120,000's!") to confirm the accuracy of
my location. Strewn garbage and the leavings of homeless camps littered the
sandy trail into the degraded pine forest remnant. The hammering of roofers
and carpenters drowned out the chipping of cardinals and the scolding
buzzes of Carolina wrens as we left the pavement behind us. Background
noise told us that Radio Road was channeling its relentless flow of automo-
biles to thousands of unknown destinations. Number 44's radio signal beeped
steadily away, but now, hours after I located it, the collar was transmitting
from a different location. Suddenly the gag had become a little too complex
for even my crafty coworkers.

Taking a northerly fork at a downy rosemyrtle (another noxious exotic
shrub from the tropics), I froze in my tracks as I stared at the unmistakable
pug marks in the sand. Too large for a bobcat and lacking the toenail marks
and symmetry of a dog track, these prints were clearly those of a male
Florida panther. I had visions of finding a half-eaten corpse of a homeless
person stashed in a palmetto thicket in this 100-yard-wide strip of woods in
developing Naples. We followed his tracks (occasionally bisected by dog
tracks) northward along sandy construction roads to the south edge of Radio
Road. Waiting for a break between vehicles, we ran across four lanes of asphalt to the brick and wood-paneled entrance where carpenters were putting
on the finishing touches. At this point the pounding of the radio collar indicated we were very close to it. The intermittent pulse rate, activated by move-

ment, confirmed what I now knew—this was no joke. Number 44's tracks resumed at the edge of a colorful flower bed before disappearing behind the fence no more than 20 yards from the road. This was where the wayward panther would spend the rest of the day before resuming a week-long odyssey through industrial parks and spreading housing developments of eastern Naples. Palmetto clumps, pines, and melaleuca provided what little cover existed.

What made this location all the more amazing was that between 10:30 A.M. and 12:30 P.M.—the time it took to land the plane and get to the site—Number 44 had traveled in broad daylight through a busy part of town. My best guess is that he crossed Radio Road before noon, shortly after I had first located him. We asked the fence builders if they had seen anything unusual in the last couple of hours, but they made no mention of large, long-tailed beasts. In fact, during the week that Number 44 stayed in town, we received no reports of panther sightings from anywhere close to Naples and found no evidence that he was feeding on local pets. By 29 April he had returned to the forests of Belle Meade, undoubtedly a lucky but much wiser cat for his troubles.

With this bizarre chapter of his dispersal complete, Number 44 began another gradual shift to the east before eluding our attempts to keep in contact with him. Because we incorrectly assumed that his travels would return to the more productive forests of northern Collier County, it took half a week before Darrell finally located him in the last place we expected: Everglades National Park. The park had once been home to what appeared to be a thriving subpopulation in this mostly treeless but extensive tract of public land. By the time Number 44 was born, however, the Everglades population was effectively extinct. Its survivors, including Female 23, had all migrated northwest to the Monument Unit of the Big Cypress National Preserve, a large, open pine and cypress forest in eastern Collier County. The Monument Unit, once referred to as Raccoon Point, was part of over 690,000 acres of National Park Service land that had been devoid of a resident panther population for nearly ten years. Even before the early 1980s there was no evidence that this area was important to panthers. Nonetheless, this zone of poor habitat would have a profound effect on Number 44.

To grasp the influence of landscape and other panthers on Male 44, we must travel back in time five years to the other side of the state. Efforts to capture and radio-collar panthers in Everglades National Park began in the fall of 1986, after NPS biologist Sonny Bass had lobbied for years to initiate a study similar to ours. Until this time, park superintendent Jack Moorehead had opposed panther research because, as best as I could tell, it was aesthetically offensive to him. The thought of a panther wearing a high-tech collar being spotted by a visitor was nearly a universal nightmare among NPS administrators. Biologically, of course, this position was indefensible—and besides, there were over 1 million acres of "wilderness" in the park. (In fact, only a small percentage of this vast landscape contained the kinds of upland forested habitat that panthers prefer, but prior to 1986 no one knew how panthers used the Everglades.) And if a 120-pound lion could roam the streets of Naples without being seen, what were the odds of anyone seeing one in this huge park? Nonetheless, Sonny's years of record-keeping and observations suggested the cats were there, especially in the remnant pinelands of the Atlantic Coastal Ridge near the park's research center. Most of the rest of this tropical pine and hardwood forest now lies under a blanket of asphalt and concrete stretching from Homestead to Jupiter—no doubt the reason why southeast Florida is so unimportant to panthers today. The Atlantic Coastal Ridge must have teemed with panthers and other wildlife before its productive forests were cleared.

To satisfy state and federal requirements regarding endangered species, a skeleton version of the GFC's capture crew was needed during the November and December 1986 capture efforts to allow the National Park Service to obtain its study animals. The first month passed without any luck, though we did come across the tracks of an adult female on several occasions. After scouting for panther sign in the early morning hours from a three-wheeled all-terrain cycle, I spent most afternoons updating Lovett Williams' 1978 account of the panther for *Florida's Endangered Biota* series. Roy would get out hours before daylight and follow his hounds across pinnacle rock pinelands and sawgrass marshes around the research center. This landscape was certainly one of the harshest and most unforgiving of any I had experienced. Although it is widely regarded as a haven for wetlands-associated wildlife species such as white ibis, woodstorks, and American alligators, the Everglades has never been thought of as good habitat for upland wildlife species. Even domesticated mammals had their problems. The hounds seemed predisposed to a host of complications that included snakebites, unknown lethal pathogens, and hungry gators. Water or matted vegetation often covered deep holes with protruding blades of limestone on every surface of the hidden rock. Travel through marshes was made difficult by deep, claylike marl, then made uncomfortable by neck-high sawgrass that slashed skin and frayed garments. Mosquitoes, heat, humidity, few roads, and relentless sun completed our Everglades experience.

7 December 1986 ■ While the rest of us were forcing down early morning cups of coffee or pretending that a pop tart was something tasty and nutritious, Roy treed the first panther in Everglades National Park. His call came through clearly on our walkie-talkies, and within minutes we had converged on a gated trail originating from the research center road. By monitoring the radio collars of the hounds we drove to within 400 yards of the tree. Barking led us the rest of the way through the waning darkness and across the hazardous Swiss cheese of Everglades pinnacle rock.

My first view of the panther was spectacular: the animal and Caribbean slash pine silhouetted against the dark purple of dawn (Figure 11.1). The eastern horizon had begun to glow, backlighting the panther in its perch 30 feet up in the first fork of the tree. Suddenly even mosquitoes and poisonwood seemed tolerable. All of us felt a wave of relief as a month of frustration lifted. Sonny's joy, no doubt, immeasurably exceeded everyone else's.

Even before the 80-pound female was in our hands, I had the impression we were looking at an animal subtly different from any panther we had yet encountered. As the sun rose, a thick, bushy, *straight* tail was clearly visible as it dangled from the horizontal limb. The first dart hit and within minutes the cat fell squarely onto the crash bag. Other differences became evident with the panther on the ground, especially the concave look of her facial features. Her fur exhibited a lusher, softer texture. Number 14 was thoroughly examined, deemed fit and healthy, radio-collared, and released—heralding in the official study of panthers in the Everglades.

By early March 1987, another adult female (Number 15) and a subadult male (Number 16) had been captured and instrumented. The resident male, the sire of Number 16, was never caught, though, and his sign was never seen. These captures and the subtly different physical characteristics exhibited by all Everglades panthers created a stir among everyone involved in panther recovery. Analyses coordinated by Melody Roelke revealed that these new study animals were genetically different as well. While the exact course of events leading to the presence of these animals will never be known, apparently they originated from introductions into the park two decades earlier of animals that were born and raised in captivity by Lester and Bill Piper at Bonita Springs Everglades Wonder Garden—a roadside tourist attraction established by the Piper brothers four decades before the capture of Number 14. The Pipers' original animals were captured in the wilds of Hendry County in the early 1940s, and from then on their offspring were ostensibly advertised as Florida panthers. Although the written record of the Pipers' captive pedigree and releases is sketchy, a female panther of unknown origin became part of their breeding nucleus. The introduced animals apparently had a genetic blueprint composed mostly of Florida stock but included ancestry of unknown geographic origin. Later speculation would claim South America as the most likely source of the wayward genes. I doubt we will ever know the true origin of this particular cat.

Figure 11.1
Female 14 was the first panther captured in Everglades National Park and looked very different from the panthers we had come to know in southwest Florida. (Note the straight, bushy tail.)

The likelihood of a genetically tainted population of panthers in Florida soon sparked a debate. How should we handle this unexpected turn of events? An extreme proposal recommended the eradication of the Everglades panthers in order to prevent further erosion of the true Florida genotype. It seemed to me, though, that the debate was pointless because we will never know what the "original" Florida panther looked like. Genetic analyses were impossible for the original Piper captive stock—or any other panther born before 1980—so the argument was moot. In the end, nothing came of the debate. These genetically altered animals were left to demonstrate how they had readapted to their harsh environment.

The discovery of the "new" panthers in Everglades National Park stimulated discussion on the original distribution of Florida panthers and their place within the larger framework of the cougar species. Before the spread of huge urban centers and sprawling agriculture, panthers in south Florida were no doubt linked by means of extensive natural forest connections. Panthers could move east and west across southern peninsular Florida along the south shore of Lake Okeechobee through cypress and custard apple swamps now converted to sugarcane production. These were the same sloughs and thickets that effectively hid the Seminole Indians from Union troops in the late nineteenth century. Panthers could also cross the peninsula on the north side of the lake through marshes, swamps, and prairies now converted to cattle pasture. Perhaps the most important link to the north was eliminated in the 1880s when the Caloosahatchee River was dredged past its headwaters in Hendry County. Lake Flirt, just east of La Belle, was as far upstream as a canoe could paddle from the Gulf of Mexico. Beyond, to Lake Hicpochee and Lake Okeechobee, stretched 12 miles of uninterrupted pine flatwoods, hardwood hammocks, cypress swamp, and freshwater marshes. Traces of the original Caloosahatchee River are still apparent as a lifeless trickle in aerial photographs and topographic maps of the area. The clearing of forests, intensive agriculture, busy highways, and artificial waterways such as the Caloosahatchee and Kissimmee rivers may have effectively isolated panthers and eliminated their historical travel ways. Any panthers left in southeastern Florida would appear to be even more isolated due to the influence of the nearly treeless Everglades.

Given the panther's tendency to avoid unforested terrain, it seemed unlikely they would traverse the open marshes dividing the Everglades and Big Cypress Swamp. In my mind the Shark River, a broad, treeless slough stretching from U.S. 41 to Florida Bay, presented a formidable barrier to the movements of a large terrestrial carnivore. For several years after the first panthers were captured in the park, my assumption was borne out by Sonny's daily monitoring of the Everglades panthers. The instrumented females produced several litters over the next four years. And except for the forced incest due to Male 16's close kinship with the resident females, this small population appeared surprisingly robust. The health and productivity of Everglades panthers were a positive and unexpected bonus for panther recovery. But this rosy picture was to change abruptly before the fifth year of study was complete.

By the end of 1991, panthers were effectively extinct in the park. Several losses were attributed to mercury poisoning. But as Sonny and I pointed out in the summer 1991 issue of *National Geographic Research and Exploration,* the general lack of forest cover made this area only marginally suitable as panther habitat. Moreover, the drought conditions that had prevailed for several years before their demise may have created a drier than normal landscape that encouraged local population growth in an area that was often under water. Panthers preferred uplands, and Everglades National Park was mostly treeless wetlands. Prevailing drought conditions may have encouraged the release of mercury into the environment—thus making what appeared to be better habitat a toxic panther cul-de-sac. Sonny and I concluded that even under the best conditions, the Everglades was not ideal panther habitat. Thus we could expect panthers to appear and disappear as environmental conditions fluctuate.

The three survivors of the Everglades die-off dealt with the event in two ways. Adult Male 16 remained in his home range—alone and presumably searching for a mate. The other two abandoned the park, crossed the Shark River "barrier," and entered the eastern Big Cypress National Preserve. This delayed double-dispersal event was unusual enough by itself, but it seemed to reinforce the idea that the park lacked some vital aspect of panther habitat. True, we had seen home range shifts due to the disappearance of adults in neighboring home ranges. But social ecology did not seem to be the driving force here. Because Number 42 was relatively young, his movements might be described as typical subadult dispersal. Number 23, on the other hand, was five years old and in her reproductive prime. (She had also spent a fair amount of time in captivity after being abandoned as a kitten by her mother.) Dispersal in large carnivores is typically a process in which young animals leave occupied range. In these cases, however, both panthers had escaped from a landscape incapable of supporting a permanent population. Numbers 23 and 42 had left nothing behind.

As Number 44 grew up under the tutelage of his mother in the productive forests of private ranchlands, Numbers 23 and 42 established unusually large home ranges alongside Female 38, an adult captured in Raccoon Point during February 1990. Number 42 would incorporate both females into his travel circuit, but only Female 23 would bear his young. Thus with only Numbers 23 and 42 engaged in reproduction, National Park Service lands supported an effective population size of two panthers on almost 2 million acres. Even though it boasted more forested upland habitat than Everglades National Park, the southern Big Cypress Swamp was still not ideal panther range. Poor-quality soils, often inundated for months, limit greatly both the prey production and the dense undergrowth that panthers prefer. Nevertheless, it was better panther habitat than in the Everglades.

9 February 1993 ■ By now Number 44 was closing in on the scene of recent panther extinctions in the Everglades. Since leaving his birthplace he had covered over 150 miles as the crow flies and avoided intolerant resident males while seeking a home range of his own. His appearance in Everglades National Park surprised everyone who had followed radio-collared panther movements (Figure 11.2). Number 44 was the first panther documented to leave productive southwest Florida for this marsh-dominated park. Now the link between the Big Cypress and Everglades was proved to work in both directions. Number 44 arrived several years too late, however, to encounter resident females. Instead he found a vast wetland ecosystem nearly devoid of

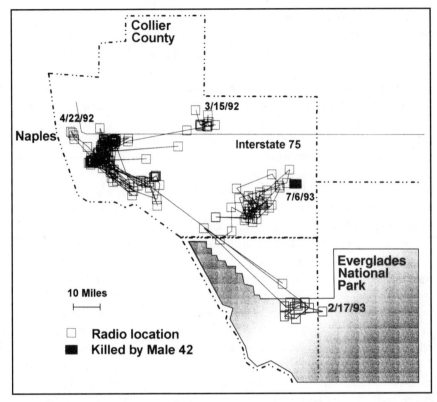

Figure 11.2
Dispersal movements of Panther 44. Male 44 surprised everyone with his foray into Naples and dispersal into Everglades National Park.

panthers. Only Number 16 occasionally passed through this part of panther range. During his two-month stay in the park Number 44 spent most of his time on Rattlesnake Ridge, a mountainous ledge of limestone towering just inches above the waters of Shark River Slough.

17 February 1993 ■ Number 44's temporary residence in the park presented us with a logistical dilemma. Although he was due for a collar replacement (to allow for growth after his 1992 capture), he now occupied a roadless and inaccessible wilderness. Fortunately Sonny was able to obtain the services of the park's helicopter to transport the capture crew, dogs and all, to Number 44's vicinity. Now that Melody was no longer with the GFC, I was lucky to have enlisted John Lanier, a Naples veterinarian with a small-animal practice, as a volunteer. Although all the pieces were in place to pull off this recapture, I did not look forward to chasing a panther through water and sawgrass while fighting off the flu. But John Lanier and the helicopter both happened to be available on this one particular day, so I would have to go into the field complete with a little dehydration, lightheadedness, and body aches.

Sonny knew the terrain, so I asked him to accompany Roy and the hounds while I looked for aspirin. John Lanier and I, with several packs of equipment, went on the helicopter's second trip. By the time we arrived, the cat was treed but he jumped and ran before we could get going. Roy had left the dartgun in the marsh near the landing site. ("Dave, it's sitting right next to a clump of sawgrass. You can't miss it.") So we spent the next half hour looking for the gun's brown case in a sea of brown grass. The day was clear and warm—about 80°F—and with the lack of a vegetative canopy on the Hurricane Andrew–ripped hammocks, the cat was surely hot.

After recovering the dartgun amid the dry, beige sawgrass, the helicopter returned and carried us to the edge of the cat's hammock where I put the gun together and passed it off to Sonny. I told Roy to go ahead and shoot before we tried to get any closer because the cat was only 6 feet off the ground in a broken gumbo limbo tree. Amazingly, he stayed in the tree after darting and in five minutes was ready to fall. We had the net under him and dogs in hand when Number 44 lost his balance—and got hung up by his head on a branch. Jayde scrambled up the slippery, red-peeling bark and quickly reached the cat—but the dead limb he stood on snapped and both panther and Jayde fell as one onto the net and Sonny below. With such a short fall the scene was merely comical, not dangerous, as human and panther limbs intertwined inside the net. The panther's temperature was only 105°F and the workup went quickly and smoothly. He had a ripped toe on a front foot from some previous encounter, but otherwise he looked well fed and healthy. Within thirty minutes we began shuttling back to the Homestead Airport—which appeared more battered by the recent hurricane than the self-regenerating tropical hammocks of the park. As the helicopter skimmed over remote sawgrass prairies and isolated tree islands, I wondered how much further Number 44 would go. Would he link the panther's functioning west coast population with its disappearing east coast remnant?

From this point on, however, Number 44's future would be increasingly affected by well-intended but shortsighted managers. Although Number 44 was remarkable as a wide-ranging and occasionally hard-to-find panther, little else was unusual about him. When he wasn't busy redefining the limits of panther dispersal patterns, this average-size male engaged in normal day-to-day panther activities. Killing deer and seeking shelter from the elements were likely his biggest concerns. The stimulus that drove Number 44 out of the park during the early summer of 1993 is anyone's guess, but the lack of conspecifics, especially females, tops my list of possibilities. Despite their solitary nature, the panther's social interactions are the fabric that weaves individuals together into a population. All our dispersing males seemed content to live alone at the fringes of occupied range—but only for a few months. Eventually they all returned to areas of permanent panther occupation with a sense of confidence that seemed to grow with their strength, weight, and age. Perhaps like humans, these nearly adult panthers experience the flashes of immortality that are so much a part of a teenager's psyche. And just like human adolescents, young male panthers seem to lack the good sense necessary to survive. Some of the antics I remember from high school are as hard to explain as a two-year-old male panther confronting an eight-year-old resident male over the rights to an adult female. Although the fabric of human society is usually flexible enough to absorb the shock waves caused by impulsive teenagers, such is not the case with panther society.

The results of direct interactions between panther combatants were brutally consistent: almost without exception, the younger, less-experienced challenger lost. The price for lingering around sexually receptive females was death to the young panther. Crushed skulls, broken femurs, deep punctures, and bacterial infections resulting from fighting accounted for eight of twelve deaths of young male panthers between 1987 and 1993. But given the lack of females in Everglades National Park, Male 44 and the lone adult male, Number 16, had no reason to fight and probably did not encounter each other during their two months of potential overlap. Lack of females was also likely the reason for Number 44's abandonment of the park. It certainly was not for lack of food—he was fully grown and in excellent condition when we changed his collar. Number 16's fidelity to an area totally devoid of female panthers is, on the other hand, inexplicable.

A significant event for Number 44 was our intervention into the life of resident adult Female 38. By 1993, three years after her first capture, this large, healthy-appearing panther had failed to reproduce. Her frequent male companion, Number 42, had sired two litters with Female 23, so it seemed reasonable to believe that the problem resided within Female 38 herself. Discussions at interagency meetings led to a plan to take Female 38 to Gainesville where reproductive experts could examine, diagnose, and possibly treat the problem. This was not an unprecedented decision. In 1987 we had removed another panther under exactly the same circumstances. Number 8 had not reproduced in the three years she was studied as a resident of the southern Fakahatchee Strand. Repeated encounters with the potent Male 12 and her failure to conceive led to the decision to capture Female 8 and transport her to Gainesville where she became the focus of great attention in captivity. She was never returned to the wild. An unlimited diet resulted in slight weight gains, but repeated medical examinations failed to discover or correct her reproductive deficiency. (She died in captivity at approximately fifteen years of age on 20 August 1988, apparently of old age.) Within weeks of Female 8's disappearance from the wild, Female 9 began a gradual shift into the home range of the older panther. Subsequently Female 9 produced her largest litter in Female 8's former range. And this was the result we were hoping for in the eastern Big Cypress Swamp where reproduction was such a rarity.

Number 38's extraction from the Big Cypress National Preserve was fairly straightforward. She treed in a low laurel oak, was darted, and was carried to a helicopter. Our new veterinarian Mike Dunbar, who flew out with the cat, was flabbergasted when he disembarked from the aircraft at Oasis Ranger Station and witnessed a passionate Farewell to Number 38 Committee composed of dozens of NPS employees. Clearly her departure struck emotional chords with many of the preserve's professional staff, a fact that would complicate management of the female panther's situation. There was an unabashed desire to have Number 38 returned to the preserve as quickly as possible, even if her reproductive problem could not be fixed.

My animated conversations with preserve staff failed to convince them that a barren female served no constructive use to the population. If she remained in the wild, unable to conceive, I argued, Number 38 would simply be taking up space that might support a healthy, contributing female member of panther society. Even if another female did not fill her place, her neighbor, Female 23, might increase her own reproductive output by having more resources available to her. In this sense, I argued if Number 38 could not be treated, she would actually become a population depressant. Further, in the absence of pregnancy, she became sexually receptive about once every month, resulting in encounters with the resident Male 42. Unable to recognize the futility of his actions, he became just as aggressive, lustful, and territorial as if each encounter was his last. The only outcome from these empty liaisons would be injury or death to a competitor over breeding rights to Number 38. Thus, I asked, what possible usefulness did a sterile Number 38 provide to the panther population and its recovery? To my utter amazement they had an answer: "Number 42 *enjoyed* Number 38's company!"

I had hoped to create some consensus among field staff from philosophically diverse agencies. I had hoped that, based on biology and a common goal of increasing the population of panthers, we would make the right decision. In this case, an unanalyzed and unrepaired Number 38 should have remained in captivity where clinical research could be conducted on hormonal or other chemical imbalances affecting ovulation and conception. Or she could have been used to help develop artificial reproductive methods and augment the captive population. But emotions and ignorance of panther social ecology (despite the several papers on the subject) prevented us from taking advantage of this management opportunity. Number 38's stay in captivity lasted only long enough for several invasive tests. Intensive surgical examinations revealed no obvious obstruction, malformation, or other physiological abnormality that could explain her infertility. Thus, after a brief recovery period, she was returned to her home range just as sterile as when she left.

Inevitably, Male 44 was drawn toward Female 38 like a moth to a campfire. For several months we watched the development of a ménage à trois, Numbers 38, 42, and 44 orbiting around each other in the remote pinelands north of Oasis. After Number 44 sent us a false alarm of his premature end, I began to doubt my own dire predictions. Maybe aggression was not so inevitable between the two males.

12 April 1993 ▪ Darrell radioed to us from the Cessna that Number 44's radio collar was in mortality mode several miles north of Oasis Ranger Station in eastern Big Cypress National Preserve. (Our transmitters had to sit still for two hours before the mortality signal was activated.) Jayde and I loaded up our two-seat buggy and hurried to the preserve. We were expecting to recover a bloated, maggot-infested carcass.

After bouncing 4 to 5 miles down a decades-worn buggy trail, we finally detected the faint radio signal of the panther. Although the signal was no longer in mortality mode, radio collars on recently dead animals are often jostled by scavenging vultures that reset the electronics to the active mode. So we continued south through towering, open pinelands and around smoothly sculptured cypress domes before finally locating the signal in a tropical hammock. At this point Number 44 heard us, got up, and walked away.

7 July 1993 ▪ Our next alarm came one month after Female 38's return to her home range—just long enough to get back into her monthly reproductive (though sterile) cycle. This situation was different, however. This time Numbers 38 and 42 were both in close proximity to Number 44's transmitter—sending its unmistakable, incessant mortality signal. Darrell and Jayde departed for the preserve hoping to find Number 44 rejuvenated just as before. Instead, they found him in a decomposed state, picked apart by buzzards. His tail and entrails were missing and he apparently had been dragged 30 or 40 yards into an open marsh. Numbers 42 and 38 were both nearby. These were all classic signs of a panther death caused by aggression induced by a sexually receptive female.

Had Female 38 not been returned to the wild (or had she been capable of reproducing and been with kittens), Number 44 would likely have moved through this recently colonized area and Number 42 would not have responded so violently to a chance encounter with a younger male. Instead a combination of natural behavior repeated over millennia, and the more recent development of human management, brought a fascinating chapter of modern panther ecology to its perhaps inevitable and violent end. Panther 44 had braved the burgeoning growth of Naples, traversed vacant panther range in western Collier County, slipped through the ranges of three resident adult males, and visited an Everglades National Park devoid of panthers before succumbing to the most common form of panther mortality: another panther.

LIVING IN THE 12
ENVELOPE

Although panthers such as Number 44 were usually unsuccessful in their crusades to establish territories, they kept our work interesting and taught us to expect the unexpected. The social pressures within the breeding core of the panther population continually forced the annual crop of young males to the peripheries of occupied range. These were the individuals who, under conditions that prevailed centuries ago, would have linked south Florida with other cougar populations to the north and west. On more than one occasion we were teased by panthers that flirted with breaking the bonds of south Florida, and on one occasion a panther from south central Florida may have been looking for a way in. Whatever the cause for the wanderlust of young male panthers, it is a phenomenon common to all cougar populations. Luck, bad or otherwise, often dictated the fate of these befanged propagules—spreading across the landscape without a predictable pattern. And luck, by creating the opportunities to study them, was often our only ally. With each young male we captured and studied, it became apparent that south Florida was like an envelope—inelastic, hermetic, sealed by a century of land use changes.

29 November 1988 ■ The dream-shattering ring of the Gainesville Econo Lodge bedside telephone jolted me out of a deep sleep. Fumbling in the dark for the phone, I imagined all kinds of family tragedies involving my wife, Diane, and our children Clif and Erin. Diane was under the distinct impression that family illnesses and natural disasters always happened at home in Collier County when I was out of town. So it was with extreme relief that I received the news from the GFC's Lakeland Regional Office. It was only a panther that had been hit by a car in Lee County near the Southwest Regional Airport.

I was in Gainesville to attend a Florida Panther Technical Advisory Council meeting at the GFC's Wildlife Research Lab. Walt, Darrell, and Jayde were in Naples to continue our ongoing studies and respond to emergencies such as these. Following my abruptly interrupted sleep, I fought to gain a balance between my cobweb-filled mind and adrenaline-enriched blood. Angie, the GFC's graveyard-shift radio dispatcher that night, brought me up to date. Commission Lieutenant Catherine Kelly was at the outskirts of the airport at Fort Myers with an injured panther lying near the road. Too far away to assist, I rudely awakened Walt by phone and turned the emergency over to him. He called Darrell, Jayde, and Robert Dilbone (a local veterinarian), then hurried to recover the injured panther. Apparently, the cat, had collided with a slow-moving airport security patrol car and lay stunned on the road before limping into a thicket of Brazilian pepper at the edge of a small tract of pine flatwoods. A combination of the immobilizing drugs Ketamine and Rompun was administered with the dartgun so the animal could be placed in a dog box inside Darrell's enclosed pickup truck. Darrell and Walt then made the four-and-a-half-hour trip to Gainesville and met me at the University of Florida's School of Veterinary Medicine later that morning.

The panther, a young male approximately eighteen months of age, sat in the crate designed for Roy's hounds and stared calmly through expanded-metal walls. Veterinary professors Jack Gaskin, George Kolias, and Elliot Jacobson met us at the small-animal clinic and helped us transfer their new patient into the facility. Suddenly the young panther found himself in even stranger surroundings, with polished ceramic tiles and stainless steel replacing his home of cypress swamps and live oak forests. By five o'clock that afternoon, George Kolias called me at the GFC's lab to report that the panther had a broken right scapula (shoulder blade) and a broken right zygomatic arch (eye socket). Remarkably he had sustained no internal, pelvic, rib cage, or long bone injuries, but he did have a slight skull fracture.

30 November 1988 ▪ Before leaving for south Florida I took a moment to look in on the injured panther. He was alert and feisty, but there was swelling around his right eye. X-rays revealed the broken bones, open epiphyses (growing ends of long bones and proof of his youth), a tail kink, and the bony remains of an armadillo, his last meal in the wild. I thought it would be just a matter of time before we put him back into southern Lee County, where no panther had been tracked before.

Over the next few weeks we received encouraging news on Number 28's progress while we kept track of his relatives back home. In south Florida there was no shortage of distractions for us: Number 12's radio transmitter had failed after his bout with Number 25 and a busy capture season was upon us. I left for the Mountain Lion Workshop in Prescott, Arizona, a gathering composed mostly of researchers and managers dealing with cougars living west of the Mississippi and experiencing distinctly different political pressures on their work. As the meeting had a hunting-oriented focus, the reports from Florida stood out from the other presentations on the species as a game animal. Our discussions were of benefit to all. But what I looked forward to most on this trip was a week with mountain lion biologist Kerry Murphy and his research associates in Yellowstone National Park.

A week with Kerry, Jay Tischendorf, and John Murmane at 6000-foot elevation and −20°F temperatures was quite a contrast to my study in the subtropics. We used Kerry's "pot-licking" hounds to hunt a cougar family group, took measurements from a male lion killed outside Yellowstone by a hunter from New Jersey, took tissue samples from an elk that had been killed by a female lion, learned to identify lion tracks in deep snow, marveled at stumps still burning from last September's forest fires, and gaped at the overwhelming scenery. I was mistaken in my fantasy, though, that many years of handball playing and running had prepared me for the rugged conditions of Rocky Mountain lion habitat. In fact, no amount of Florida training could acclimate me to this strange and rarefied environment, and I constantly suffered the effects of a stressed-out respiratory system. High altitude, deep snow, and fleet-footed mountain men quickly wore down my sea-level subtropical conditioning. But overshadowing the body aches, headaches, and cold was a memorable day spent with Jay Tischendorf and his borrowed hound.

12 December 1988 ▪ Jay and I hunted for lion sign in the Bear Creek drainage, a mountain system that supported gold mining and lions on private lands outside Yellowstone. We found tracks of a bobcat and a female lion on a snow-covered mountain road overlooking Gardiner, Montana. Jay released his single dog (a beagle named Abby) on the trail of the lion, but only after the bobcat's tracks had veered off in single file downslope. Abby picked up the female lion's scent immediately, and we found ourselves in pursuit of the barking, short-legged beagle as she plowed through the snow. The dog trailed the lion for about five minutes, then chased her for another two before treeing her 40 feet up a snow-laden spruce. Jay's shot with an antique dartgun was excellent, but it caused the cat to leap to another tree like an acrobat. She then sprang back to the original capture tree before hurling herself down a wooded draw. I tried to trail her while Jay repacked his gear, but I lost her tracks amidst a confusion of elk and mule deer sign on the ski trail across the creek. Abby, however, raced ahead of us and found her again as she stumbled across an open hillside then slid another 20 yards to the flat terrace of the ski trail. We estimated her age at seven to nine years. She was in excellent condition and lactating. Her fur was incredibly lush, and her fat, furry tail had the diameter of my forearm. By comparison, the typical Florida panther's tail was like that of a rat—thin and file-like.

Jay's new cougar was the product of an environment with its own unique hazards. Florida's mosquitoes and humidity were replaced in Yellowstone with porcupines and dry deadly cold. The local climate, no doubt, explained many of the differences between this mountain cougar and the ones I knew in Florida. Yet I was reminded of the bushy-tailed panther in the Everglades—the similarities between these two cats were striking. Perhaps the true bloodlines of Female 14 and her kin in the Everglades had closer genetic links to the American West than the origins in South America that had been suggested by others. After the half-hour workup, conducted with numbed fingers, we attached a radio collar and left the 103-pound mother under the protective boughs of a low-growing spruce.

What sticks in my mind about this capture in rugged, snow-covered mountains is the simplicity and efficiency of the operation. Though Jay and I enjoyed each other's company, I was just a bystander. Jay could just as easily have conducted the capture alone: he was prepared to handle the hound, carry all the equipment if necessary, anesthetize the lion, draw blood samples, affix the transmitter, and, after her recovery, track her movements. What a switch from our captures of Florida panthers: hundreds of pounds of equipment, off-road vehicles, redundancy in personnel, an intricate division of labor, and usually balmy weather. As we watched Jay's newly collared female recover, I realized that during our panther captures in Florida, my most important function was ringmaster of a three-ring circus, not field biologist. The complexity of panther captures—spawned by the accidental 1983 death of the decrepit female in the Fakahatchee discussed earlier—was a result of agency anxiety about image. Capture mortality is always a possibility when wild animals are handled, and adding a handful of field technicians to what should have been a simple procedure made our operation no safer for them. It was truly a joy to participate in a good old-fashioned lion capture. I spent most of the next day pondering these contrasts in Mammoth Hot Springs as the hot volcanic waters soothed aching muscles and ice crystals formed in my hair.

The rest of my week at Yellowstone was quite memorable—soreness and exhaustion, great scenic beauty, biting cold, vast numbers of elk, bison, and mule deer, and by the last days a feeling of acclimatization to the cold and altitude. On a hike to an elk calf kill we saw two coyotes coursing across the road and through a snow-covered expanse. They stopped every dozen steps or so to turn and watch us watching them. They appeared quite suited to the environment (unlike some of their scruffy immigrant cousins now living throughout Florida). We also saw a lone coyote run up a steep embankment near the Boiling River on the way to Mammoth Hot Springs—a dipper was using a footbridge for cover near the parking lot while a black-billed magpie engaged in acrobatics on a juniper across the road. This was a place where wildlife research could be conducted in a site of unparalleled beauty, where the individual was still the main unit in fieldwork, where political distractions seemed far away and of little consequence. Of course, Yellowstone is not immune to controversy. The Craighead brothers, Frank and John, had pioneered the study of grizzly bears during the 1960s before being banished from Yellowstone as researchers because their management recommendations clashed with National Park Service policy. And today Yellowstone is embroiled in the management and politics of the reintroduced gray wolf— an icon of western environmentalists but a bane to sheep and cattle ranchers.

*　*　*

After my glimpse into traditional mountain lion biology, I returned to the strange milieu of Florida panther politics. In January 1989 we captured new panthers Numbers 29, 30, 31, and 32 and recaptured Panthers 12, 17, and 18. But as Number 28 steadily healed in confinement in Gainesville, backstage activity began to cast a shadow of doubt on his return to the wild. The Florida Panther Population Viability Workshop—sponsored by the U.S. Fish and Wildlife Service and conducted by the Captive Breeding Specialists Group—had created an atmosphere of panic. As many panthers as possible must immediately come into captivity. Because Number 28 was already in Gainesville, the case was made to keep him there as a permanent ward of the state. I argued that it was more important to give him the opportunity to open new windows into the southwest Florida panther landscape. As the first panther captured in Lee County, Number 28 came from a beleaguered system of dwindling forests under tremendous development pressure. The Audubon Society's Corkscrew Swamp was the heart of this twisting wetland labyrinth. Its capacity to support panthers and its connection to the rest of southwest Florida's panther range depended on a narrow corridor known as Camp Keais Strand—a swamp that was shrinking in its vital functions like the plaque-clogged arteries of a human cardiac patient. Ever-expanding agriculture and residential development figured to be the embolism that would shunt the movement of panthers through this ecological artery. Along with a new state university and airport expansion targeted for the north end of this forest subsystem would come the inevitable network of roads, services, and housing developments. A better understanding of panther habitat use in this area could help Lee County planners direct future development in a more environmentally sensitive manner. Number 28 could be an important new experiment—if the agencies were willing to try.

Late in January 1989, however, Melody Roelke announced that a potentially disastrous disease, feline infectious peritonitis (FIP), had been isolated in a colony of western cougars housed in Gainesville and destined for experimental release into north Florida. Number 28 was being housed at the same facility. If Number 28 were returned to the wild now, Melody argued, he could spread the disease throughout south Florida. Thus Number 28 might be destined for a life in captivity regardless of what the agencies wanted. And the possibility of FIP infection might keep Number 28 from teaching us important lessons in landscape ecology that could benefit all wildlife in southwest Florida. By early February, however, veterinary researcher Jack Gaskin had cultured the fungus creating the symptoms erroneously interpreted as FIP. It was an organism unique to the Sonoran Desert Life Zone and posed no threat of persisting in dank, humid Florida. Apparently it had arrived with one of the Texas cougars. Despite the serious doubt cast on the likelihood of Number 28 transmitting FIP to Florida wildlife, Melody continued to worry about the disease because it had decimated a colony of captive cheetahs some years ago in Oregon. In the final analysis, this "medical emergency" faded from memory like so many others. In the meantime, a healed Number 28 awaited his fate.

A month and a half later, Number 28 was to be released in the National Audubon Society's Corkscrew Swamp Sanctuary. My efforts to have him returned to one of the private ranches close to the Southwest Florida Regional Airport where he was found were stymied when ranch owners refused to allow us access to their property. Corkscrew Swamp, although 12 miles away from the airport, was part of the same wetland system through which Number 28 had been traveling when he was hit. More important, the National Audubon Society and the sanctuary's superintendent, Ed Carlson, welcomed the rehabilitated panther with eager anticipation.

15 March 1989 ■ Darrell and I left Gainesville at midnight with the panther loaded into a wooden crate. We felt like we were making a clandestine prison break for a member of our gang. Earlier in the day we had drugged Number 28 to load him and refit his radio collar. For the next five hours, the panther retraced his journey of nearly four months earlier. We arrived at Corkscrew Swamp at 5:00 A.M. and slept for about an hour under the thatched roof of a sanctuary chickee hut (a traditional open-sided Seminole Indian structure that finds popular usage throughout South Florida). The media showed up en masse as well as a few agency personnel on a south Florida junket. In all, thirty-five people watched Number 28's release as I began to pine for the minimalist conditions of Yellowstone. The door to the crate was opened at 7:45 A.M. on Little Corkscrew Island. Number 28 peered out of the crate, looked left and right, then bounded twice through dry marsh grasses before trotting north into a live-oak hammock. The brief and anticlimactic event was followed by a short burst of applause from an audience mostly unaware of the tribulations leading to his return but appreciative of the chance of seeing him.

*   *   *

The winter of 1988–1989 marked a turning point for me—the point where I began to think that political concerns were outweighing the needs of the panther. I sensed there would be little time left to complete our field studies before political decisions prevented us from learning what we needed to know about the south Florida landscape and how to keep the panther a vital part of it.

Previous to Number 28's accident, nothing was known about panther distribution in this rapidly developing area. As a result of this absence of data, the GFC's Office of Environmental Services, which reviews development plans, had effectively written off Lee County as panther habitat. Although we assisted in their reviews by evaluating the potential for panthers to occur on a particular site based on vegetation that appeared suitable for the big cats, our opinions carried little weight without tangible evidence. My brief review of a planned 1000-acre development known ironically as "The Habitat" was typical of the reviews most new projects received. In a 28 February 1986 letter to Geza Wass de Czega, the biological consultant on the project, I wrote:

> This letter is in response to your request for a survey of *The Habitat,* for the DRI (Development of Regional Impact) of this proposed residential development in south Lee County. While we have not documented any evidence of Florida panther use in the area of the DRI, we feel that it is very likely that some degree of use occurs here. Because of the nature of vegetation and proximity to known occupied panther habitat, it is my opinion that the land in and around *The Habitat* is suitable for this species. In addition, National Audubon Society's Corkscrew Swamp Sanctuary, which is adjacent and contains similar vegetation components, produces regular sightings and other evidence of regular use.

This letter, however, was insufficient grounds for our agency to mount meaningful opposition to the project. Two months later it was approved by the GFC's Vero Beach office.

Three years after the GFC's initial review of the project, a radio-collared Number 28 wandered through the site and confirmed our original suspicions. Although this documentation of Number 28 did not prevent the issuance of construction permits for this development, its design would be transformed to minimize environmental impacts on wetlands and wildlife including panthers. The permit issued by the U.S Army Corps of Engineers for the filling of wetlands would require the developer to leave significant areas of forest on the property that would adjoin neighboring conservation lands. Number 28 had, like no other panther before him, affected the future of the southwest Florida landscape.

For several weeks after his release, Number 28 did nothing remarkable in terms of panther behavior. His locations centered on Corkscrew Swamp Sanctuary, especially within the hardwood hammocks of Little Corkscrew Island. Gradually, however, we found him farther and farther from the sanctuary, until his movements took on the characteristics of a ping-pong ball bouncing across a landscape-sized table. Over the next four months we found him less than a mile north of Alligator Alley near the Collier County landfill, in the second-growth swamp forest of lower Corkscrew Swamp, at the edge of a new development in Lee County known ironically as Wildcat Run, at the north end of the Lehigh Acres airport, within a mile of my house in northern Golden Gate Estates, and, finally, on the Baucom Ranch, a forested island positioned between the Caloosahatchee and Orange rivers within view of the Lee County power generation facility just outside of Fort Myers.

During Number 28's travels I experienced some of the most interesting situations I had encountered in our panther research. Maybe, we joked among ourselves, this young male's unusual life experiences were the result of his head injuries. Considering his youth and the geographic context of his dispersal, the real surprise is that Number 28 did not get into bigger trouble. During the first four months of his repatriation, he could rarely have been out of earshot of such busy highways as I-75, State Roads 80 and 82, and County Roads 951 and 846. Human activity was a continual part of his environment. Yet throughout this time we received only one report that he had been seen by anyone.

11 August 1989 ■ Dick Highmark, the Immokalee vegetable grower/ bobcat hunter, the man who developed the knack for treeing our study animals, called our office at around 4:30 P.M. to tell me a rambling "hypothetical" story of a dog-eating panther at a home on the west side of Lake Trafford in northern Collier County. His description of the residence was very near Number 28's location that day. For the last month this panther's movements had been limited to this area, and I had been wondering if he was injured.

Highmark, who spoke in a hesitant fashion, hard to follow, finally turned his tale of "what if" to nonfiction. The homeowner, Jack Queen, a local construction contractor who had a young panther treed in his back yard, had seen the panther on his back porch after he heard his dog barking hysterically. Apparently, when he turned on his porch light and peered out his back door, he was greeted with the sight of his dog suspended in the vise-like grip of a pair of panther jaws. Queen's sudden presence must have startled the cat, for it then dropped the dog. The panther then casually walked away. The next morning I visited Queen and examined the dog that now had a puncture on the upper left side of its cranium and another under its right ear. Miraculously, the small dog was not only alive but ambulatory. Because Number 28's telemetry location on the day of the reported incident had been in the vicinity, we assumed it was him, even though Queen did not remember seeing a collar. At a distance of 4 feet, however, the image of his dangling dog must have overridden a myriad of other details. Queen's only memories of the cat were that it was "light brown, had a long tail, and it was *huge*." No doubt he had seen a panther, and no doubt it was Number 28.

During his rambling dispersal, Number 28 passed by the site of his original accident and capture at least four times, suggesting that his presence there was no fluke. (Four years later another young male was hit and killed near the same spot on 9 August 1993.) With the exception of his encounter at the Queens', Number 28 left no wake even though his circuitous forays took him through suburban developments along the outskirts of Fort Myers, Naples, Immokalee, and La Belle. Apparently this was a cat destined to test the limits of the panther envelope along with his own survival abilities. One month after his interrupted canine snack, he crossed State Road 29 just north of Immokalee (and less than a mile from Dick Highmark's house), entered Hendry County, and wound up in the thicket swamp of Lake Hicpochee, a wide spot in the now-channelized Caloosahatchee River. It seemed we were on the verge of witnessing the connection between panthers living in the Big Cypress Swamp and the remnant, hypothetical population inhabiting an array of private ranches in Glades and Highlands counties. The modern history of panthers in this disjunct suburb of the Big Cypress included the old female that Roy had treed in 1973, a male that was roadkilled on 19 March 1983, and a full-grown male tracked only with sign by Jayde from April 1984 to May 1985. It was with great interest, then, that I listened to Dr. James N. Layne of Archbold Biological Station in Highlands County recount the discovery of "steaming" panther sign in a large patch of scrub vegetation. Jim and coworker Doug Wassmer had found tracks and a warm scat containing hog remains on the grounds of the station, a private research facility that lies about 7 miles south of Lake Placid. Discovered at 8 P.M. in an area of flatwoods and seasonal wetlands, the tracks lead purposefully along a sandy jeep trail before disappearing into the woods.

This was the first physical evidence of panther occupation from this part of the state in nearly three years. With the 1988 capture season looming ahead, Jayde and I traveled to Lake Placid, met with Dr. Layne, and set up a sign survey in and around the area of the mystery cat's first appearance. During November, Jayde found male tracks on many of the wooded ranches near the biological station, but, interestingly, this cat was never documented again at Archbold. Almost like clockwork, though, his sign appeared in abundance about every two weeks on one of the ranches where we had obtained permission to look. Unlike many of the ranchers in Collier and Hendry counties, the ranchers in Highlands County were mostly happy to permit our searches. By the time Roy and his dogs returned from Texas, I had planned a busy capture schedule that allowed only two weeks to hunt for panthers in the southern terminus of the Lake Wales Ridge.

We used the well-appointed station as our base of operations during the first week of searching. Each day we would split up to scour the trails winding through bayheads and scrub or bordering the many citrus groves covering this ancient ocean dune line. In the evening we would engage in rugged games of basketball in the Archbold parking lot before turning in early for the next day's hunt. Although the panther's tracks were ubiquitous, we could not find sign fresh enough for the hounds to do their job. He even left tracks as a long trail of pug marks through a small citrus grove—but nothing suggested that he lingered in this impoverished wildlife habitat. His tracks were those of a cat striding purposefully, but unhurriedly, to get from one side of the grove to the dense bayhead on the other.

Driving our balloon-tired swamp buggies through hilly, arid scrub terrain—instead of the mud and inundated sands of the Big Cypress Swamp—was a strange sensation. Gopher tortoises appeared frequently in our path as they searched for succulent herbs at the edges of trails, and scrub jays would interrupt their morning foraging to peer down at our strange vehicles as we passed by thickets of sand hickory, rosemary, and Chapman's oak. If nothing else, we all enjoyed the change in scenery.

For the last week of the search we left Jayde and Roy to pursue the mystery cat by themselves while the rest of us returned to Hendry County to conduct panther surveys in Gum Swamp. My plan was for Jayde to summon us through our regional radio system if and when they caught the cat. We would then beat a hasty trail to the tree. The final week passed without finding any sign more encouraging than we'd found the first week. By now, at least, we knew that our quarry's diet consisted mostly of wild hog and white-tailed deer. In addition to tracks, he had left scats as clue cards on this scavenger hunt.

30 January 1988 ▪ Roy and Jayde volunteered to spend this Saturday hunting for the panther. This would be the last day of our search. We would need Sunday to prepare for a long-scheduled Monday capture to replace Number 12's aging transmitter, and we could no longer divert our efforts away from south Florida. Like the rest of us I was at home in Collier County, assuming that our wait was futile, when the telephone rang at 11 A.M. I was in no hurry to answer it. After all, it was too late in the day to expect the dogs to be working effectively. But our radio dispatcher had an urgent message to relay: Roy needed us at the Hendrie Ranch just east of U.S. 27 near Venus, Florida. Within minutes of the call I had most of the crew rounded up and we were a fast, green caravan speeding through southwest Florida's checkerboard of farms, ranches, citrus groves, and remnant wilderness on our way to the northernmost capture site of a Florida panther. We arrived in Venus (now just a lonely post office at a wide spot in the highway) just two hours after the call. Fifteen minutes later we were gazing up at one of the most beautiful panthers we had ever captured.

As I had suspected, Roy and Jayde had indeed given up their quest at around 10 A.M. when the temperature rose and the wind picked up. The dogs were loaded into their crate and the buggy was weaving through dry scrub sands when Roy spotted a fresh scrape in the middle of the road. Remarkably the dogs indicated considerable interest in this sign even though they had been at work since before sunrise. Within moments they were eagerly trailing the panther that had eluded us for two weeks. Now the three-to-four-year-old male was treed in the lowest fork of a large slash pine and looked quite comfortable in the shade of the tree. If not for the panther's double-crooked tail and a backdrop of swamp bays, the scene could have been a snapshot of a western cougar in an Arizona ponderosa pine forest. After two weeks of teasing us with his sign and frustrating us with his elusiveness, Number 24 cooperated fully during his capture. With 20 feet of tree still above the cat, we inflated the crash bag in case the dart persuaded him to climb higher. His graceful tumble into the cushion was followed by a routine and speedy workup. As the stars appeared on Saturday night we packed up our gear and speculated on where this panther would take us.

One of the special pleasures this capture conferred was the opportunity to share it with Jim Layne, one of the discoverers of Number 24 and active in panther recovery efforts since the 1970s. Though Jim did his best to maintain his scientific detachment, no force could have masked his joy of seeing an animal so rare but in his opinion so inadequately studied outside the Big Cypress region. To Jim, Number 24 proved that not enough had been done to look for panthers in an important part of their range. He believed many other panthers lived in the patchwork of ranches in Highlands County and beyond. Perhaps he was right. But our mandate in southwest Florida didn't allow us the time to search in this area.

Number 24 opened a new window into panther distribution and habitat use, but he did not give us the conclusive evidence of other panthers that Jim Layne was so confident of. Perhaps the panther's young age excluded him from the resident females whose territories would be vigorously defended by older resident males. Or perhaps we just did not track his movements long enough to find the places where other panthers resided. In any case, our panther work north of the Caloosahatchee River was short-lived. Except for those of us in the field, there was apparently no interest for continued searches for panthers in a large area dominated by private property. Looking back, I'm surprised we even had the chance to add Number 24 to our study. And I remained troubled by the official denial that panthers *might* exist outside the Big Cypress and the enclosures of captivity. Effectively, agency administrators had decided for the people of Florida that Glades and Highlands counties were wild enough and the panther had no part in their future. This tacit declaration seemed inconsistent with Rowdy McBride's 1989–1991 search for panthers in Arkansas that was funded, in part, by the U.S. Fish and Wildlife Service.

The kind of data we got from Number 24 would have been cause for great celebration by those scouring the hinterlands of Colorado for sign of the elusive grizzly bear. For decades, dedicated volunteers and a handful of agency personnel have been looking for irrefutable evidence of a Rocky Mountain bear population that may well not exist. Even so, Colorado's Division of Wildlife has conducted a long-term study that resulted in the capture of scores of black bears while researchers attempted to document the presence of their larger cousin. It would have taken much less effort to scour the remaining wildlands of Florida.

Number 24 provided us with only seven months of data, which we published in a paper in the *Florida Field Naturalist*. In general, his use of habitat was similar to the patterns we had noted in southwest Florida. Although he preferred upland forests over wetlands and agricultural lands, Number 24 occasionally used scrub and bay forests, as well, vegetation unavailable to panthers farther south. His wide-ranging movements covering nearly 400 square miles took him as far south as Palmdale and as far north as the outskirts of Avon Park, a distance of about 40 miles. Highlands Hammock State Park, near Sebring, was a one-day stopover for him as he explored northern

Highlands County. All of his other radio locations were on private property. Apart from the different geographical context, Number 24's travels were similar to the dispersal movements of Number 28. And, like Number 28, his birthplace was unknown.

Given the large, albeit fragmented, amount of forest cover that remained in south central Florida, the proximity of radio-collared panthers, the historical records of panthers in this area, and recent verified sign on the Arbuckle Wildlife Management Area, we concluded that Glades and Highlands counties were in fact important to the recovery of the Florida panther. This area could offer a region for natural expansion of the southern Florida population through dispersal as well as reintroduction. But expanding landscape conservation measures was not an administrative priority.

In mid-August 1988, Number 24's radio transmitter began emitting an unusually fast signal—stuck midway between the normal active mode and the faster emergency pulse rate. Without Roy in Florida, we would have to wait another five months before replacing the collar. Unfortunately, we would not get that chance. On 22 August 1988, Jayde and I found Number 24's decomposed carcass about 1 mile east of U.S. 27 beyond a grove of planted eucalyptus trees on the Lykes Brothers' ranch in Glades County. His transmitter switch had failed to activate the mortality alarm. Only a week's worth of unchanged radio locations had suggested something was awry. Our experience with Number 24 ended where his life had probably begun—in a dense palmetto thicket. Maggots had already initiated his return to the soil leaving few forensic clues to his early demise. And not only his death remains a mystery. We may never know if this part of Florida supports breeding panthers or is simply a dispersal sink where young males end their explorations.

Panther 28 failed to make the short swim into Glades County and connect his movements with those of the late Number 24. Instead he turned south and set up a fairly predictable home range in southern Hendry County ranch land where we had few radio-collared adult females but a surplus of young adult males. During the 1991 capture season, we had the first opportunity to see Number 28 since his release at Corkscrew Swamp.

7 January 1991 ■ Number 28 appeared to be close to the previous day's location, and Number 29 was a little farther east of him. As I sat in my truck before the sun came up, Number 28's signal was so close that it activated my radio. Roy and I approached the panther with the two dogs Squirt and Amy at about 6:45 A.M. Amy jumped him almost immediately. She treed Number 28 in about ten minutes, but Squirt trailed some other cat (maybe a bobcat), barked, and distracted Amy from the treed panther. To the dogs, I suppose, the possibility of an as-yet-untreed cat was better than a dozen panthers already up a tree. Roy and I searched for the cat in a hammock adjacent to a cypress dome. We knew he was close because we could pick up his signal without our antennas. Finally I saw the cat about 4 feet off the ground and less than 20 feet away. We stared at each other, face to face, for a moment. But before Roy could exit a small grove of cabbage palms, Number 28 jumped out and ran off. Within fifteen minutes Roy had called the dogs back and the chase was on again. Number 28 retreed in about five minutes, this time in a low live oak. Everyone else arrived twenty minutes later.

Balanced in a branch-studded live oak, Jayde dropped Number 28 into the net after an injection of the newest immobilizing drug, Telozol, and three minutes of waiting. Overall he appeared healthy, but his front half was covered with scars, punctures, and other minor injuries. His right front foot was swollen and there was a fresh puncture on his right wrist. He also sported a shallow puncture on top of his head. His rear half was remarkably free of wounds, perhaps a sign that he was the aggressor in violent encounters. He weighed 121 pounds and had a 107°F fever, possibly due to his foot infection. We repeated the electroejaculation procedure, now standard on all male captures, and froze eight pellets of his sperm in liquid nitrogen for the future. We rated his sperm quality moderate: we saw some vigorous cells on a slide under high magnification.

24 January 1992 ■ Number 28's next capture came almost two years after he had killed Females 18 and 41—the disruption discussed earlier as the probable result of the natural demise of Male 17. Directives from Tallahassee had increased our workload this year by requiring recaptures of the collared adult males on an annual basis (versus the standard two years) in order to obtain more sperm samples for cryopreservation. Although I had objected to increasing the frequency of capturing study animals—due to the hazards of anesthesia—Number 28 was targeted for recapture after just one year.

Darrell located Number 28 on the Big Cypress Seminole Indian Reservation. Roy and I went after him on our small two-seat swamp buggy but had a hard time homing in on him because of all the commercial interference on his frequency. (The radio channels used for wildlife studies are not unique and sometimes a collar or two will be transmitting over a crowded frequency band.) We followed the dogs Jody and Susie through palmettos, melaleuca, and cypress before Number 28 treed in a low, scrubby live oak in the middle of a palmetto thicket surrounded by large pines.

Because he was so low in the tree and looked inclined to jump, I left most of the team behind and just used the net with the minimum crew of four people. After darting, he stayed in the tree and hung up in its branches. Jayde climbed and easily lowered him out—at which point Melody began shouting that Number 28 was not breathing well. Suddenly tempers were short. Everything had seemed to be going so well. Perhaps the presence of a *National Geographic Magazine* photographer had something to do with the higher-than-usual anxiety. But Number 28 did appear to stop breathing for a minute or so. In fact, this was as close to a panther dying as I had seen in seven years—an important reminder that, despite all our precautions, captures of wild panthers would always be unpredictable and hazardous operations. Panther behavior, panther condition, environmental conditions—these are variables that cannot be controlled, and any one of them can interact with drugs and human error to cause a panther's death. It was for this reason that we began deploying collars designed to transmit for three years. We hoped they would remain in operation for that time period without the need for a potentially dangerous capture. After all, the aim of wildlife telemetry should be to minimize disturbance of study animals and produce good unbiased data. Captures should be viewed as necessary risks that are conducted only when absolutely essential.

At least we had eliminated a troublesome radio collar. Number 28's right foot appeared to have been broken a while ago and it had a very large leading toe, which made a distinctive pugmark in mud and wet sand. He now weighed 128 pounds and exhibited an increasing peppering of white flecks. After giving him the usual battery of injections, we fitted him with a new radio collar. Aside from his torn and tattered ears, he reminded me of Number 12 at his first capture. Number 28 had become that rare male who survived to adulthood.

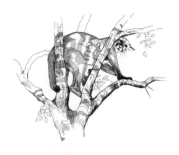

THE PANTHER'S EDEN

13

There is a place in northern Collier County, at the western edge of the Okaloacoochee Slough, where herds of thirty or more deer and sizable flocks of wild turkeys can be seen feeding near the edges of large swamps and hammocks. It is a quiet place, receiving much less use than the publicly owned Bear Island with which it shares a common boundary. Within this 16-square-mile ranch is a large cypress swamp that provides habitat for otters, barred owls, and pileated woodpeckers; extensive hardwood hammocks that provide abundant food and cover for wild hogs and armadillos; interconnected freshwater marshes and prairies dotted with egrets, cranes, ibises, and wood storks; and scattered pine flatwoods that support fox squirrels, swallow-tailed kites, and diamond-back rattlesnakes. The high diversity of plant communities, the productive soils, and conducive drainage conditions make this a haven for wildlife. Moreover, its owner is jealously possessive about his landscape jewel—keeping it under lock and key save for a very few confidants and his immediate family. Little, if any, hunting occurs here and it is patrolled closely to keep trespassers out.

Panthers inhabit this Eden. In fact, they have used it to such an extent that over an eight-year period rarely a week could go by without our recording multiple locations of radio-collared study animals. Here a panther could pick and choose each meal, selecting from a menu more diverse and well stocked than anywhere else in Florida. It was here that Female 19 conceived and raised her recordsetting first litter, where adult Males 12, 13, and 20 jockeyed with each other for breeding rights to resident females, where Number 11 raised at least four litters. Probably more kittens have been reared on this plot than on any other comparably sized area in Florida, perhaps in the United States.

These incredible demographic statistics are even more remarkable when viewed in a landscape context. Fully 50 percent of the property has supported some form of intensive land use since before oil was first discovered in Sunniland in 1943. Vegetable farming, citrus production, and rock mining combined to alter a once impenetrable wilderness. But in the last few years development has jumped: a convenience store advertising gas, ice, and cold beer has been erected by the owner at the ranch's northwest corner, vegetable fields have expanded, pines and palmettos have been cleared, and preparations have been made to expand the rock pits.

To grasp the pace at which this area has developed, it is helpful to note that virtually none of Collier County had experienced either the plow or pavement in the year 1900. By 1973, the land that is now the Sunniland Ranch was supporting more than a square mile of intense agricultural and extractive uses. In addition, an extensive system of ditches had been installed to drain the property's wetlands. By the time we began tracking panthers on the ranch in the late 1980s, these land uses had more than doubled their 1973 coverage. Today the alterations continue, and the land is now open enough to support coyotes. One can only imagine the level of panther use on this tract of land before the changes began.

The pattern of use by panthers, however, has been erratic. A peak in radio locations was recorded in 1987 when relatively few animals were radio-instrumented. This peak resulted from a concentration of activity surrounding Female 11's family, Female 19's early maturation, and Male 13's attraction to potential breeding opportunities. No doubt the cyclical nature of kitten production has influenced the roller-coaster documentation of panther use on the ranch. But an overall decline has occurred: annual totals of radio locations began with 406 in 1987 and dropped to 77 in 1993. Unless one infers that the very act of monitoring radio collars caused this decline, the only reasonable explanation is that panthers have responded to the loss of excellent habitat on the property. The pattern that continues to unfold on the Sunniland Ranch is the same pattern that has eliminated panthers from 95 percent of their former range. And it has happened with little leadership from government agencies to stop it.

It was clear to me early on that this was a special place. The owner, Florida native Jack Price, would grudgingly allow me to visit the property on occasion to investigate the activities of our study animals. Price was a large and forceful man who launched readily into diatribes against government in general and the National Park Service in particular, even though he was under contract with the Big Cypress National Preserve to remove the fill associated with abandoned oil rigs throughout the preserve. But his dealings with this agency and his witness of the "ruin" of Bear Island under NPS stewardship had left a bitter taste in his mouth. The ruin, in his view, stemmed from the improper management of the preserve—too many people using it, too little prescribed fire to maintain vegetative vigor, too little local experience to maintain its quality.

At one point in 1988, Price afforded me the rare distinction of temporarily giving me a key to the ranch. Behind the locked gates I had a number of memorable experiences. My encounter with Female 19 and her kittens was paramount, but there were others. During Number 19's precocious first courtship I had the opportunity to show a panther to my eight-year-old son Clif. We were waiting at the edge of a large palmetto thicket as Darrell walked in on the cat while cardinals and brown thrashers flitted in and out of the thicket's dense cover. Clif tugged on my sleeve, pointing excitedly, while the young adult female exited the thick vegetation and briefly showed herself in a clearing. Several months later, after Number 19 surprised us with her first litter, Jayde and I searched a dry cypress swamp for sign of her family. Near the site of her morning location was an excavated hole at the base of a tree surrounded by a palette of kitten tracks. Just a few paces away were the remains of their earlier attentions—a neatly dissected armadillo carcass. Not a bone or shred of flesh could be seen within the calcareous exterior of this tanklike exotic mammal. The shell had been mashed flat as if it were a platter with two handles made of the armored top of its head and bony tail. Every digestible morsel on the half shell had been consumed, but not before the kittens had practiced their hunting skills on their hapless quarry.

These visits to the ranch came to an end when the Game and Fresh Water Fish Commission rejected Price's precedent-setting plans to build a convenience store where no commercial development existed for miles around. This opposition seemed consistent with the GFC's philosophy and the needs of the panther, but Price held me guilty by association. He felt it was wrong for us to oppose the granting of permits when he had afforded us access to his property for research. I tried to convince him that the GFC's opposition would have occurred with or without his cooperation. Any other decision, I said, would have been professionally irresponsible. But from this point on, even though the store was built, we were unwelcome on the ranch.

Our telemetry flights, however, provided an ample view into the dynamics of land use versus panther use on the property. It became increasingly clear that our panthers' Eden was being gradually reduced to a much tamer version of the verdant paradise we first encountered in 1986. Expansion of improved pasture, vegetable fields, and citrus groves was making the panther's preferred habitat and den cover less remote and more islandlike. Panthers continued to use the ranch, but their presence was on the wane.

*   *   *

Certainly the loss of forest cover was the primary cause for localized panther abandonments, but there was more to this than simple agricultural and urban expansion. Government agencies can be extremely good at developing geocentric management plans for a particular tract of public land—that is, they can focus quite well on land within the boundaries of their surveyed and posted preserve. But this narrow view of the world assumes that large preserves are sufficient to maintain the increasingly threatened wildlife species that live on them.

Panthers have an uncanny way of exposing the inadequacy of most public preserves in supporting even a single individual of the species. What the panther shows is that private lands are essential to the continued existence of large cats in Florida. Nearly every Florida panther ever captured has used private land—some nearly to the exclusion of public property. Yes, the Panther Habitat Preservation Plan has been formulated and a set of maps produced. But today, more than three years after its adoption, no progress has been made in enlisting private landowners in the fight to save the Florida panther. Part of this problem has been caused by the inability of agencies to deal with more than one landowner at a time. And on the rare occasion when they are dealt with as a group, private landowners can be quite intimidating, as we eventually found out.

At a public hearing at the old Lee County courthouse in Fort Myers on 4 February 1994, the last week of my employment with the GFC, I listened mutely as officials representing the Florida Panther Interagency Committee (FPIC) attempted to explain to an unruly audience the blessings of the Florida Panther Habitat Preservation Plan—the document born of the lawsuit brought against the U.S. Fish and Wildlife Service by the Fund for Animals three years earlier. The plan was intended to identify those parts of south Florida that were currently occupied by panthers or might someday support them through natural colonization or reintroduction. The plan also recognized that some of these lands might be occupied by panthers even though their status there was unknown. In addition, the plan outlined several methods by which this array of private land, which totaled about 1 million acres, could be acquired or at least protected. In preparing the maps that were to worry the landowners, Todd Logan and I had spent many hours poring over unfurled aerial photographs, satellite images, and topographic maps, looking for the largest patches of forest and the most logical connections between them. We ignored landownership boundaries and examined only current land uses. At the end of our mapmaking process, a lengthy text—composed by agency administrators—was added. Then, almost as an afterthought, it was shared with a few members of the agricultural and development communities.

Perhaps the responses of FPIC members to pointed questions from anxious property owners were influenced by the stress and discomfort of being put on the spot. But I suspect they were unable to explain what the Habitat Preservation Plan was all about because they had so little appreciation for the intricacies of panther biology and complexities of panther landscape. For more than an hour agency policymakers Buck Thackeray (National Park Service), Jack Pons (Department of Environmental Protection), Dennis Jordan (FWS), Tom Logan (GFC), and Al Egbert (GFC), failed to assure the audience that they were only seeking cooperation. Time and again the agitated crowd was told that the thick tome outlining the vast areas needed for habitat protection was nothing more than a planning document—despite the fact that the U.S. Fish and Wildlife Service had hired a full-time employee to implement it. The response that epitomized the meeting, and perhaps panther recovery in general, came after an exasperated farmer had his question read by the meeting's moderator: "What land uses does the plan find compatible and what land use is not compatible [with panthers]?"

Tom Logan's response was almost immediate: "Everything that Jack Price has done on his land is compatible." This was followed at once by an unidentified shout from the audience: "Can you put that in writing?" In these brief sentences Tom Logan had undermined thirteen years of research conducted under his administration—research that clearly pointed to the dangers that development poses to the future of the Florida panther. Literally every problem threatening the existence of panthers—roadkills, inbreeding, natural disasters, limited dispersal range, frequent intraspecific aggression—was created or exacerbated by precisely the kinds of land uses that were so quickly isolating Jack Price's panther Eden from the rest of south Florida. The ranch continues to support panthers, not just because the management of its dwindling forests favors them, but because the remaining good habitat is nearby larger tracts of occupied range. Comparably sized patches of forest exist elsewhere, but they are isolated from larger forests and no longer support panthers. The Sunniland Ranch itself cannot support panthers—but, ironically, public lands intended to support them must continue to do so if the panther is to survive even on better-quality private land. In other words, from the perspective of long-range panther conservation, public lands need private lands as much as the reverse.

The agencies could not admit that a tactical error had been made in preparing the Panther Habitat Preservation Plan. Many of the audience's concerns came from landowners asking why they were not consulted in developing a document that could have such a profound effect on them. If nothing else, landowners would have had some stake in the process if they had been consulted earlier. But instead of simply admitting that the process was imperfect from the outset, the agency representatives steadfastly defended the plan by stressing that it was only a planning document and participation was voluntary. A local corporate attorney summed up the audience's anxieties. If a landowner did not voluntarily participate in the plan, the attorney suggested, then the regulatory agencies would make the permitting of agricultural and residential developments as costly and time consuming as possible in order to extract from the landowner what volunteerism could not.

Landowners must *want* to help in the recovery of the Florida panther. Coercion will not generate authentic support. Many of them would like to help, but few can afford it. In order to pay their taxes and maintain their quality of life, property owners must make their land produce income. If it is truly agreed that forests on private land must be part of the future panther landscape, then their owners deserve to be compensated for sacrificing income or the potential to earn income on their land. What we really needed was the appointment of a panther habitat lobbyist—someone knowledgeable about panthers, south Florida, ranch owners, and local politics—who could develop proposals to local, state, and federal governments that would provide tax incentives, cash, or other compensation for the voluntary protection of key habitat. Such a program is not only feasible, it is necessary.

This, I believe, is the crux of the panther's plight. Extinction will result, not from poaching or roadkills or inbreeding or disease or competition with hunters, but from habitat loss. Given the pattern of landownership in Florida, it is clear that the private sector holds all the keys to panther recovery. The inexorable chipping away at habitat by private landowners and the inaction of natural resource agencies are a dangerous combination—more damaging to the panther's future than an open season on this endangered cat.

Even more troubling than the indications that wildlife agencies don't know what they should be doing is the possibility that they know exactly what they are doing. In his book *Twilight of the Panther,* Ken Alvarez devotes over 500 pages to branding panther recovery efforts as inept and counterproductive. He observes:

> Most officials who maneuver with such assiduous determination to make a nonsense of recovery programs act from selfless motives. Though the enactments may be aided and abetted here and

there by personal ambition, ignorance and incompetence, they are driven in the main by the loftiest (and most uncompromising) ideals. This may help to explain the curious situation in Florida in which four government agencies responsible for planning the panther's survival fashioned a compromise recovery plan without making any rigorous inquiry into its prospects for actually working. The principals were each looking primarily to see that it did not work against their established interests.

It is doubtful that such a detailed and depressing account of wildlife policy will be written again soon. It could, however, help others to avoid drafting similar blueprints for disaster. The meeting in Fort Myers did nothing but reinforce Alvarez's condemnation of those in control of the fate of Florida wildlife. Sadly, he drew numerous parallels between panther recovery and other failing or failed conservation efforts such as the dusky seaside sparrow. The patterns were not unique to the Florida panther.

Ironically, many owners of large ranches in south Florida are proud to have panthers on their land. Not once, in nearly nine years of dealing with ranchers in Collier and Hendry counties, did I hear disparaging remarks about panthers killing livestock or eating too many deer. But when landowners perceive that overregulation threatens their livelihood, then suddenly the panther becomes a costly liability and its future on private land is cast in doubt. Still, to most landowners, the panther is one more reason their ranches are special places.

*   *   *

What would happen if we were to lose all remaining panther habitat on private land? Such a loss would reduce the population to barely twenty—an exceedingly high estimate because it is based on the assumption that all panther habitat is of equal quality. Even if all panther habitat on public land was of the highest quality, I doubt it could support more than a handful of them. Unfortunately, the outcome of the Fort Myers meeting suggested that only public lands would one day support panthers.

Even before the unveiling of the Habitat Preservation Plan at Fort Myers in 1994, agency administrators felt fairly uneasy about the prospects of keeping private land as part of the south Florida panther landscape. In fact, another recovery approach was under consideration. The notion of genetic restoration was not new to panther recovery, but it did not receive serious consideration until a Florida Panther Interagency Meeting was convened at White Oak Plantation in 1992. In attendance were members of the Captive Breeding Specialists Group, the organization that just a few years before had recommended the removal of panthers to start a captive breeding program. Using data and generating results that were similar to the 1989 species survival analysis described previously, this group concluded that the panther's genetic status, presumably eroding, could be fixed by directly translocating wild cougars from as close to south Florida as possible. After the meeting at White Oak, plans to accomplish this artificial interchange were slowly drafted. After the unveiling of the Habitat Preservation Plan at Fort Myers, the active pursuit of the new strategy began.

There were few drawbacks to genetic restoration so long as one believed that the species *Felis concolor* (including the Florida panther) was distributed in recent time without interruption across the Western Hemisphere, that several subspecies intergraded from one to another, and that the behavioral, physiological, and genetic differences among cougar subspecies were insignificant compared to the benefits of this introgression. And there was another benefit of this approach: it could eliminate the need to breed panthers in captivity. The U.S. Fish and Wildlife Service had always stayed out of efforts to "save" endangered subspecies with similar cross-breeding methods—the creation of hybrids, the FWS argued, violated the Endangered Species Act. As a result, private efforts to save the obscure dusky seaside sparrow were insufficient, while an upsurge of public support for a glamorous raptor resulted in the return of a genetically reconstituted peregrine falcon to the Eastern Seaboard. But these were last stands: the former to rescue an all-male population, the latter to replace a resident population long extinct. Both efforts, however, relied on captive breeding in their attempts to restore a population with either crippled or nonexistent demographics.

There are two basic assumptions underlying the Florida panther's genetic restoration. One is that the Florida panther is genetically bankrupt. The other is that, historically, subspecies of mountain lions did not maintain distinct borders but exhibited zones of overlap as parts of a larger, continuous gradient that joined all subspecies throughout the Western Hemisphere. Thus the genetic restoration plan aims to simulate natural gene flow, or immigration, by importing mountain lions from over 1000 miles away. This is an overnight fix to a potential problem that has, arguably, taken many centuries to develop—a problem that may have as much to do with rising and falling sea level as with more recent events. The basic idea is to speed up the process of natural gene flow—a process that, even in an unfragmented landscape, is exceedingly slow.

To illustrate how much time it might take for panthers to transport their own genetic material from one end of their range to another, suppose we examine several scenarios based on subadult dispersal—the mechanism by which genes become distributed in a species. Because males, by virtue of their longer dispersal distances, are the primary movers of the population, at least six individuals, each taking an 80-mile one-way trip, would be required to transport one unique gene from the southeastern Big Cypress Swamp to Pensacola, Florida. Another six trips by six more panthers would be necessary to complete the gene's journey to east Texas, where the closest extant neighbors of the nearest non-Florida subspecies reside. Let's assume that it takes five years for a male panther to successfully breed, that they all dispersed toward Texas, and that each of these males survived. Then the minimum time needed for a one-way delivery of a unique genetic message across 1000 miles would be sixty years. From our studies, though, we know that few young males survive to reproduce. So if only half of these males survived, then an equal number of legs in this genetic relay would have to be repeated, thus stretching the delivery time beyond the century mark. In reality, a survival rate for these young males closer to 10 percent would be more likely—suggesting that nearly 700 years would be required to accomplish this feat naturally. And this assumes that the Mississippi River and other barriers would not compound the delays. Not only that, but we would have to assume that a unique south Florida gene would remain part of the genetic makeup of an individual that was the product of parents increasingly different from its original kin in the Big Cypress Swamp. And these assumptions, of course, would also be true for a reverse-direction gene transfer from Texas to Florida.

What underlies the discussion of large carnivore taxonomy is not simply that it takes centuries to move genes across space, but that a cougar from Texas has never, in all probability, interacted directly with a panther from Florida, let alone from south Florida. Further, by the time genetic material from Texas could find its way to the Big Cypress Swamp, the package would have changed so dramatically as to be indistinguishable from the standing residents. Gene flow is a much more circular process than the linear models suggest. Even in south Florida where dispersal opportunities are limited, dispersal direction appears to be random. Aspects of geology, hydrology, and vegetation all encourage a high rate of local gene transfer and discourage long-range gene flow. This is the process that creates the physical attributes of species that taxonomists have used to describe subspecies and races. Even so, the lines between subspecies are rarely distinct. In pre-European North America, the westernmost Florida panther was likely indistinguishable from the easternmost Texas cougar (just as today's Rocky Mountain puma is indistinguishable from the Montana puma along their common border near Yellowstone National Park). On the other hand, the westernmost Texas cougar and southernmost Florida panther would have exemplified (and probably still do) all the differences described by Nelson and Goldman in their definitions of *Felis concolor* subspecies. Florida panthers may be all the more unique as occupants of a peninsula. Most mainland subspecies of mountain lions intergrade with neighboring subspecies. But even before the settlement of Florida, most panthers lived in a landscape nearly surrounded by water. This natural isolation, created by the rising and falling seawaters of previous glacial epochs, predisposed them to reduced gene flow and increased the likelihood they would diverge from other subspecies.

A nearly identical set of circumstances presented itself to managers of the Tatra Mountain ibex in Czechoslovakia just a few decades ago. Managers there decided that genetic augmentation would help this small, isolated population of goatlike animals that had dropped to very low numbers. When ibex from the Sinai, which has a warm, dry climate, were introduced into the colder and wetter East European setting, the distantly related individuals readily interbred—but their hybrid offspring produced young at a different time of year. As a result, all the offspring of hybrid pairs died from adverse weather conditions and the population went extinct. This event emphasizes a phenomenon known as local adaptation—that is, individuals have evolved key physiological or behavioral characteristics that are specialized for their environment. Even though these local populations may be part of a widespread species, they are isolated enough to have developed very different genetic traits that are linked to very different environments. The climatic differences between dry, mountainous west Texas and humid, flat south Florida are every bit as great as the differences between Czechoslovakia and the Sinai.

While the hybrids produced by Texas cougars and Florida panthers may not develop bad reproductive timing, the possibilities for other problems are nearly endless. Given that panthers exhibit normal demographics and are holding their own, shouldn't this genetic experiment be implemented in captivity? Wouldn't it be better to experiment in a controlled environment than to leave the results to fate in the wild? At least we might find a purpose for the grown Florida panther kittens that are currently caged without purpose. A controlled experiment, over a few generations, would show whether or not this approach is appropriate for the wild population. Because panthers in south Florida give no indications of a rapid demise, there is plenty of time to test what happens when the Florida panther is reconstituted in captivity *before* gambling with the only known wild population.

*　*　*

Is genetic restoration as planned, then, the answer to the panther's presumed genetic problems? Is it biologically defensible to introduce individuals from mountainous, rocky, and arid terrain from over 1000 miles away into a population adapted to subtropical, sandy, sea-level conditions—a population, moreover, with a different genetic makeup? This action may well provide agency administrators with a few more restful nights by eliminating worry based on the exaggerated impacts of low numbers, inbreeding, and habitat loss. But if panthers continue to function in a demographically normal manner after Texas cougars are introduced into Florida, does this mean the experiment has succeeded? Or might this suggest that nothing was broken in the first place? The unspoken message—the real meaning in this recovery decision—is that for the Florida panther to be saved, it must first be destroyed.

David Ehrenfeld, past editor of the journal *Conservation Biology,* provides insight into how such decisions are made. In an editorial for *Orion* magazine (Winter 1992) he writes:

> The answers are not hard to find. It happens because of the fragmentation and bureaucratization of responsibility and the proliferation of experts, ensuring that no planner will see the entire picture. It happens because of modern, top-down management, in which isolated administrators playing their cards close to the vest make decisions for other people who will have to live with the consequences. And it happens because of the foolish infatuation of decision-makers with "state-of-the-art" systems designed with utter disregard of the needs of their users.
>
> In what was the Soviet Union, an outraged populace, fed up with decades of waste, the destruction of human cultures, and the misuse of human resources, has started to throw out some of the arrogant central planners and managers. In America, we are also faced with a glut of managers, substituting their dissociated, power-oriented fiat for the experience and judgment of the people they manage. Our universities, like our government, our industry, and our other institutions, are sorely beset with the blighting curse of imperial management, but no revolution is any closer than the distant horizon. Perhaps we are stuck with the "state-of-the-art" until the money runs out.

In two paragraphs, Ehrenfeld has summed up the process of decision making in panther recovery and explained why so few actions taken by the cumbersome coalition of agencies have resulted in positive change and why landscape approaches to panther conservation have been ignored. His essay was not written about the ineptitude of environmental bureaucracy, however. He was describing the creation of an ultramodern academic hall at Rutgers University and his dissatisfaction with the new "monstrosity." In this analogy, Ehrenfeld's monstrosity and the state-of-the-art can be translated here as panther recovery and its implementation. The lure of high-tech solutions to complex natural resource issues is so irresistible that the most practical means to a successful end are often ignored. This explains the easy entrapment and detour of panther recovery by captive breeding and the more recent derailment toward genetic restoration. As David Orr observes in the June 1994 issue of *Conservation Biology:* "The deeper problem, noted by all critics of technology from Mary Shelley and Herman Melville on, is that industrial societies are long on means but short on ends. Unable to separate can do from should do, we suffer a kind of technological immune deficiency syndrome that renders us vulnerable to whatever can be done and too weak to question what it is that we should do."

If we continue to divert funds and attention away from south Florida and its demographically functional panther population, the worst-case scenario will come true. The longer we shy away from solutions to landscape problems, the quicker we run out of management options. Now is the time to address habitat loss in places like the Sunniland Ranch and a handful of others throughout southwest and south central Florida—not after the remnant forests are so disjointed they can only support coyotes. Moreover, the government must stop characterizing the private sector as the enemy and, instead, turn to it for solutions. Private landowners must become active participants in panther recovery because they want to.

The passage of the Endangered Species Act nearly thirty years ago is a reflection of public interest in America's wildlife heritage. But, like most social programs, it has always been underfunded—in this case by virtue of government inefficiency and the depressingly large number of species in need of "recovery." Without a tax-based source of revenue to drive habitat acquisition and management, the onus of land conservation has been passed on to a very small percentage of the public. Increasingly, Florida's Game and Fresh Water Fish Commission and the U.S. Fish and Wildlife Service are pursuing land and money grants from private developers in exchange for approving their applications for development. While neither wildlife agency has legal authority to directly regulate development, their comments are taken seriously by Florida's water management districts and the U.S. Army Corps of Engineers, the grantors of prized water-use and dredge and fill permits. The process known as mitigation allows impacts to a given piece of property in exchange for, at the developer's expense, a like-sized or like-valued property of sufficient quality to be useful to some sensitive species or ecosystem.

Mitigation, as practiced in Florida should really be called "strategic withdrawal." Its original meaning implies the lessening of development's impacts—an environmental trade-off of sorts. The ultimate in south Florida wildlife mitigation would be to physically reunite the known range of the panter with the forests of Charlotte, Glades, and Highlands counties. We know that underpasses have successfully mitigated some of the influence of highway inpacts in Collier County, so why not a Caloosahatchee River overpass spanning this barrier? Plant it with palmettos and live oaks, link it with existing forest on both sides of the river, and suddenly the envelope would open and ease the pressure within the panther habitat core. Allowing the loss of panther habitat in one place should mean improving conditions for panthers somewhere else. But this is not what mitigation means for panthers and most other wildlife. In practice, agencies permit the continued loss of wildlife habitat by trading pieces of functioning ecosystems slated for development in exchange for pieces of functioning habitat elsewhere. But what is being mitigated here? Nothing is being mitigated if the land obtained by the government already contains the species of interest. It just prevents another piece from being totally lost. True mitigation can only occur if that which is lost is replaced by an equal amount of land for that species to *expand* into. Exacerbating this situation is the fact that most land given to the government to satisfy the process is undevelopable. The kinds of land dominating public preserves are classic examples. The primary reason why the National Park Service administers two of the largest tracts of land in south Florida is that their previous owners could not profitably grow tomatoes. Bad soils for vegetables will not grow productive forest cover and Florida panthers. This, if nothing else, emphasizes the need to keep panthers on the Sunniland Ranch and other private lands. Without them, the panther is like an ice cube in the desert.

MUDDLING TOWARD A SOLUTION 14

The conversion of native forest cover to intensive agricultural and urban use causes profound alterations to the landscape—alterations that will eliminate panthers if these changes are extensive. And without adequate habitat there will be no opportunity to manage either demographics or genetics. As wildlife population biologists A.R.E. Sinclair and the late Graeme Caughley observe in their 1994 wildlife ecology textbook, the most common causes of extinctions, in descending order of importance, are "contraction and modification of habitat; increased predation or hunting; competition for resources with a species new to that environment; a poison in the environment; and a disease, particularly one new to the environment." As far as panthers are concerned, poisons and disease seem to have little influence on the population. Panthers are no longer intentionally preyed upon or hunted by people, either, and the only new predator in the area, the coyote, has not yet been shown to compete with the panther. Only contraction and modification of habitat is an immediate and range-wide threat. In fact the coyote is just another symptom of habitat modification.

By and large, the most pressing need is for an effective panther insurance program on private lands. Despite a wide range of management philosophies on public lands, major changes in the extent and quality of forest cover are unlikely. But simply because land is owned by the government does not mean that its managers claim to promote panther recovery—habitat loss can occur on public land as well. The Big Cypress Seminole Indian Reservation is a case in point. Just as private property owners have not been enlisted in panther conservation efforts, neither have the Seminole Indians. Sadly, the federally administered reservation's future is every bit as uncertain as any private ranch. Although it is revered in Seminole lore and religion, the panther seems just as threatened by urbanization and expanding agriculture on the reservation as anywhere else in south Florida. But the politics of wildlife management on this large tract of public land are very different because, in practice, the Seminole Tribe is not subject to the same environmental standards as federal preserves such as Big Cypress National Preserve and Everglades National Park. There is no legislative mandate that dictates wildlife conservation on this reservation, and there seems to be little interest in ecosystem processes or their management. The Seminoles are understandably more interested in developing their economy than in making their land a nature preserve. But this also means there are opportunities here for capitalist creativity, just as on private ranches. Financial incentives will be needed to offset the lost development opportunities if panther conservation is to be effectively sold to tribal chairmen—just as incentives will be needed to convince corporate owners, trustees, stockholders, and their land managers.

One reason why so little progress has been made for panthers on private and Native American lands is that the problem appears so overwhelming to natural resource agencies. This is why debates over the biology or philosophy of panther management on public landscape—debates over minor problems such as hunting versus not hunting or the placement of a deer food plot—take precedence over the big issue of land management. This is why these agencies tend to treat symptoms such as malnutrition, parasites, and genetics rather than the problem itself. Public agencies have yet to show how the south Florida landscape can be integrated in such a way that a diversity of land management practices and ownership patterns will actually promote panther recovery.

Within the core of panther range or potential panther habitat, about a dozen ranches in south Florida hold the key to the panther's future. In Collier and Hendry counties there may be as few as six. How can we explain the presence of reproducing panthers? Clearly the ownership of an area itself is insufficient when examined alone. Many of south Florida's small preserves or private ranches, 30,000 acres or less, exhibit excellent habitat quality. Those that support panthers are adjacent to a larger preserve or form part of a complex of privately owned panther habitat. Good habitat patches without panthers are generally islands of forest that are isolated from larger forests. Thus habitat quality alone does not tell us a lot. Even property size is not particularly helpful—especially when we consider that panthers on the largest preserve, Everglades National Park, are effectively extinct and that the second largest, Big Cypress National Preserve, has fewer panthers per unit area than anywhere else in occupied range.

There is only one factor that appears to explain consistent panther presence and steady reproduction in south Florida—proximity to the habitat core. The boundaries of this area encircle a zone of forest that supports the majority of panthers. This large habitat island, or "meta-preserve," has not experienced wide fluctuations in panther demographics like lands farther south and east where local populations have gone extinct or the land is too poor to support more than a handful of individuals. Panthers denied access to this core experience a more variable environment, exhibit irregular reproduction, have much larger home ranges, and often lead shorter lives. If a forested area falls within the core boundary, then panthers live and breed there regardless of the human activities that go on. The kind of forest that makes up the core seems of much less importance than its context and connection to more forest. Panthers seem reluctant to traverse large tracts of treeless terrain—so the further an area is from this core and the more fragmented its forest becomes, the less likely are reproducing panthers to be found. Thus our largest public preserves cannot be expected to make significant contributions to panther recovery because these lands are mostly far from the habitat core and are themselves highly fragmented (Figure 14.1). Such lands are much more valuable for recreation, water conservation, and preservation of nonterrestrial wildlife species such as river otters, limpkins, and round-tailed muskrats. This is why the Fakahatchee Strand State Preserve, which consists mostly of avoided or tolerated wetland habitats, has continually supported panthers. Despite poor-quality habitat, it is an integral part of the habitat core.

The differences between occupied and unoccupied panther habitat are as noticeable from outer space as they are from the ground. Satellite imagery is increasingly being used to address wildlife management issues through geographic information systems (GIS). This technology has its roots in the military but has slowly found its way into civilian applications such as urban planning, pollution monitoring, and agriculture. It is a wonderful tool for understanding large, landscape-sized systems and the ways they have changed. This is virtually impossible to do from ground level. But the technology has its shortcomings. Although some GIS are capable of identifying land cover types smaller than 30 by 30 meters (the crown coverage of a single large live oak), you can quite literally miss seeing the forest for the trees. Nonetheless, it has been used extensively to describe aspects of wildlife distribution and habitat use. Frank and John Craighead pioneered the use of this technology for wildlife in the early 1970s while studying grizzly bears in Yellowstone National Park. The value of GIS for panther conservation is its ability to depict the variety of plant communities used by panthers and to show how the remaining south Florida forests are interconnected. Only the broadest scale is necessary to show where panther conservation efforts are most needed.

When you examine a satellite image of south Florida relative to panther

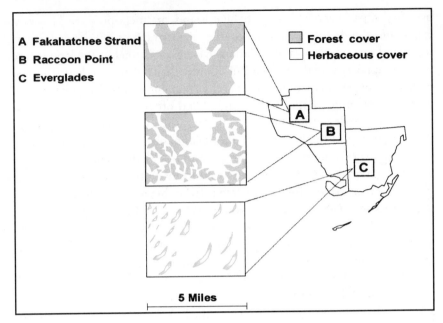

Figure 14.1
A likely explanation for low and variable panther numbers in southeast Florida is the naturally fragmented forest that characterizes this area of frequent flooding and poor soil quality. Forest cover and habitat quality follow a declining gradient from southwest Florida to southeast Florida.

distribution, a single feature stands out: a continuous green blob stretching through the Fakahatchee Strand, the Florida Panther National Wildlife Refuge, Bear Island, the Sunniland Ranch, and eastward through private property and the Big Cypress Seminole Indian Reservation to the L-28 feeder canal (Figure 14.2). To the northwest is the Corkscrew Regional Ecosystem Watershed, including Corkscrew Swamp Sanctuary, which is connected to the core by a meandering band of swampland surrounded on both sides by extensive vegetable farms. This is where Female 31 lived and raised several successful litters. Thus the privately owned Camp Keais Strand serves both as a movement corridor for panthers and as good habitat itself. Directly to the north are the headwaters of the Okaloacoochee Slough. They begin on private ranchland in Hendry County and have drainage linkages to the Caloosahatchee River further north and to the panther refuge to the south. Female 52 colonized part of this privately owned slough and gave birth to her first litter here in 1993.

East of State Road 29 and south of Alligator Alley, the forest gets spotty. The further your eyes scan to the southeast, the more abbreviated become the

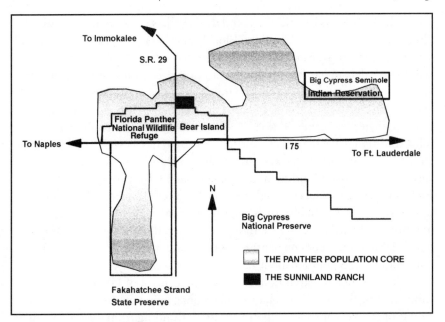

Figure 14.2
The interconnected forests of southwest Florida form the Florida panther's habitat core—
the key ingredient in the future of the population in this part of the state. A large proportion
of this key area falls on private land.

comma-shaped islands of forest, afloat in a vast sea of stunted cypress trees
and herbaceous wetlands. This is great raccoon and water moccasin country,
but it is terrible panther habitat. In fact, in keeping with modern ecological
jargon, the few panthers living in such marginal range are essentially the
"living dead" of the population. These are the population's extras, living in
areas too wet or too unproductive to contribute regularly, if at all, to the year-
to-year cycle of renewal in the panther population core. Thus the targets for
future panther conservation efforts are clear. The private forestlands in the
north and west of this habitat core must be maintained as panther habitat. To
depend solely on public lands will condemn fewer and fewer panthers to depend on more variable, lower-quality habitat.

*   *   *

Assuming that we can slow the loss of panther habitat in some areas of south
Florida and actually increase it in others, do we have the wisdom to manage
it properly? Many sportsmen and long-time residents of Collier County
claim that the joint management of Big Cypress National Preserve, and

especially Bear Island, by the GFC and the NPS has resulted in noticeable declines in both white-tailed deer and Florida panthers. They claim that the old system of private stewardship was more sensitive to natural resources and wildlife. A debate has been raging over appropriate human use in the preserve, but neither side has produced much evidence to support its argument. As a result, researchers from academia have criticized interagency efforts for not employing experimentation to solve such hurdles to panther recovery. The traditional backbone of panther research has been "natural history studies." Laboratory-style experiments, by contrast, control as many outside influences as possible and then manipulate one to determine its importance. Field studies have been essential, but they have yet to address the importance of environmental variables such as prey abundance, habitat fragmentation, and human disturbance—even though over $4 million was spent by the state and federal government on panther research and management between 1976 and 1994. Over $3 million alone has been spent since 1987.

With panther research dollars under the control of GFC and NPS administrators, there is little chance that nonagency scientists will alter the course of recovery through experimentation. Funding from the state wildlife agency barely trickles to the state's wildlife school, the University of Florida. Virtually all important research is conducted within the agencies and, unlike academic research, is not published or peer-reviewed. Today the basics of panther biology are known; the challenge is to fit the panther into the landscape of Florida's future. But the application of biological theory seems beyond the research capabilities of agency field staff and their administrators. By failing to enlist academia with its complement of internationally recognized zoologists and conservation biologists, the agencies have only hurt the chances for panther recovery.

Research is not always premeditated, however. The panther is surrounded by experiments ranging from conservation gambles (such as highway underpasses) to outright retreats from legislative mandates (such as the resistance to single-species management on National Park Service land). The answers to successful panther recovery have been presented to us like cut-glass jewels in a kaleidoscope. These jewels, like fragments of the south Florida landscape itself, can be found among the seemingly random spattering of vegetative and drainage features of south Florida. But these fragments are only decipherable if we wish to find them. When they are discovered, they provide insight into a number of panther management questions. These are the studies that should be completed before classic experimentation begins.

*   *   *

An experiment is never a failure if it adds to our understanding of a vital process. If Everglades National Park had been established to show that these million acres were capable of supporting panthers, we would grade the effort as a dismal failure. But our witness of the extinction of a reproducing panther population has taught a valuable lesson—a lesson in the importance of habitat quality to population stability.

When Everglades panther captures and telemetry studies were finally approved in 1986, the National Park Service heralded this development as an opportunity to demonstrate the benefits of a half century of limited human access for this large solitary carnivore. If that was true, then Everglades-style "ecosystem management" would become the model for panthers. For the better part of two years it appeared that this was the case. But the series of panther deaths and home range abandonments that began in 1989 reduced the Everglades population to one adult male (Number 16) who still wandered a territory devoid of females through 1994. Some agency researchers believed that mercury contamination caused the deaths of these residents, but unequivocal evidence never materialized. Even if mercury did play a part in panther mortality, most likely it only exacerbated problems for an upland species living in a mostly aquatic environment. Not only were Everglades panthers forced to use an area with little upland forest, but they frequently consumed atypical prey such as otters and alligators, species that could magnify contaminant loads in their tissues. A sparse deer herd and a lack of wild hogs increased the likelihood that panthers would eat these smaller wetland predators. Despite total protection from game hunting on this largest eastern national park, it became clear that the Everglades was incapable of sustaining even a small segment of the Florida panther population.

* * *

The National Park Service is not alone in having suggested that human disturbance limits panther numbers. The Everglades example shows clearly that low human activity alone cannot ensure panther survival. In fact, areas like Bear Island and many private ranches demonstrate that panthers will indeed tolerate some off-road vehicles, camping, and the killing of deer and hogs.

There is a place, however, where unrestricted human activity has likely reduced the population of panthers in south Florida. Since the construction of a network of canals and streets by the Gulf American Corporation in the 1960s, the southern 40,000 acres of the Golden Gate Estates have been a magnet for disruptive outdoor activities. Besides the out-of-season hunting that is considered widespread by GFC wildlife officers, our research documented the nonfatal shooting of Number 9 and the poaching of several radio-collared black bears. Almost every day people from nearby neighbor-hoods can be heard discharging firearms ranging from revolvers to machine guns. Off-road vehicle use is limited only by thick vegetation and canals.

There is nowhere else I know of in the panther's range that receives this kind of human use. Presumably this is why Golden Gate Estates is avoided by panthers despite the dominance of heavy forest cover. Further, it holds the distinction of being the only area abandoned by a resident breeding female. In this unintentional experiment, adequate forest cover was insufficient to offset the effects of widespread human activity aided and abetted by its many crisscrossed and fragmented roads and canals. The quantity of forest cover here is a hundredfold greater than in Everglades National Park. But, as in the park, prey are few and far between, and panthers are even rarer.

* * *

The Ford Vehicle Evaluation Center represents another classic experiment of panther tolerance to human intrusions. What would happen if a chain-link and barbed wire fence were built to exclude panthers from 530 acres of good habitat that also contained a vehicle test facility? The intent, of course, had been to deny panthers entry so that none would cause a dangerous acci-dent on the 2.1-mile-long high-speed straightaway. But with the continued presence of prey and a predictable schedule of human activity, the animals continued to use this exclosure. Four different panthers were found within the compound at one time or another. All of them seemed to exhibit normal movements and activity while they were there.

Less than 20 miles away Number 48 succeeded in raising her first litter in the face of intense hunting, heavy off-road vehicle traffic, and a government-prescribed fire. Again, predictable daytime human activity apparently al-lowed her to maintain a regular schedule of killing prey and raising her three kittens.

Number 48's mother, Number 31, provided additional insight into the panther's ability to tolerate people when she gave birth to a litter at the edge of a busy vegetable field in the Camp Keais corridor. Had a farmworker known that the den was there, it would have been easy to lob a cucumber right on her head. Although local human activity was occasionally intense, it was regular, it occurred during the day, and it was nonthreatening to Number 31. These examples suggest that where adequate prey and dense cover are both present, and human activities are predictable, panthers adjust or simply continue their day-to-day activities unchanged.

* * *

The logging of the Fakahatchee Strand in the 1940s created still another landscape-level experiment. How would panthers respond to the total elimination of their forest home? Would a precipitous rise in the deer population as the result of an increase in browse—and then its concomitant crash as the result of successional recovery—be their coup de grâce? Historical records indicate that throughout drastic changes in cover and prey conditions, panthers apparently inhabited the strand—first during the times of abundant prey and sparse cover that occurred immediately after logging, then in its present state with sparse prey and abundant cover following fifty years of regrowth.

The closure of the Fakahatchee Strand State Preserve to off-road vehicle use, the condemnation of hunting camps, and the creation of a no-hunting buffer on adjacent private lands in the late 1980s have had no measurable impact on this subpopulation of panthers. In fact, after thirteen years of radiotelemetry only six kittens are known to have been produced here, and none is known to have survived to adulthood. (Two males were taken into captivity as kittens in 1991 and are now part of the untapped captive breeding program.) The Fakahatchee seems to just barely provide the resources to sustain a small number of panthers, regardless of the good intent of management. No more than one breeding female has been documented to reside here at any one time. The insult of clear-cutting still seems to hang over this most unique of Florida forests.

Unless drastic measures, such as large-scale clearings, are taken to improve deer numbers in the preserve, only the healing powers of time will restore the strand to its former productivity with a high canopy of cypress and a lush understory for deer browse and panther cover. Until this happens, the Fakahatchee Strand will continue to be a population sink. But because it is a key component of the habitat core, it will continue to be occupied by panthers. Even with the temporary intrusion of loggers fifty years ago, the Fakahatchee has remained a remote and relatively hostile environment for people. No doubt the swamp's inherent resistance to humans has contributed most to the panther's persistence here despite few prey and an abundance of water.

*   *   *

The panthers' abandonment of Everglades National Park and their colonization of eastern Big Cypress National Preserve created the opportunity to measure their response to a very wet and patchy environment, as well as their reaction to an area devoid of other panthers. This area is more forested than the Everglades, but its woodlands are more open and they are fragmented by wet prairies—habitats that panthers generally avoid. By 1993 three panthers were using the Raccoon Point area of the preserve, an adult male (Number 42) and two adult females (Numbers 23 and 38). While Number 42's home range of 217 square miles was only slightly larger than that of the average male further north, the females had home ranges two to three times larger than females residing north of Alligator Alley where trees form a more continuous forest in the habitat core.

It appears, then, that lower-quality habitat and interrupted forest cover force panthers to travel more widely in order to kill sufficient prey and seek the dense daytime cover they prefer. This ultimately results in a lower density of panthers and reduces the likelihood that this small subpopulation will persist. And to make matters worse, a high water table and frequent, widespread flooding reduce further the quality of this habitat. The exchange of panthers between Everglades and Big Cypress is exactly the kind of population dynamics we should expect in such harsh conditions. More upland forest and a stable and productive environment north of Alligator Alley translate into higher panther densities and a greater probability of persistence.

*　*　*

As Everglades panthers were beginning their spiral into oblivion in 1989, Todd Logan was ushered in as the first manager of the Florida Panther National Wildlife Refuge. Todd had accepted the challenge of taking 25,000 acres of prime panther habitat and turning it into the textbook example of land management. He began by accompanying us on our research forays into the future refuge—still used for grazing cattle, harvesting cabbage palms for the landscaping business, and the pursuit of traditional outdoor recreation including camping, swamp buggy travel, and hunting of both small and large game.

As is the practice on most new federal refuges, this one was immediately closed to the public. Traditional uses quickly came to a halt. As a result, the summer of 1989 became the demarcation between the overuse of the past and the wise use of the future. Slowly, Todd increased his permanent staff with specialists trained in law enforcement, deer management, heavy equipment operation, and the use of prescribed fire for enhancing wildlife habitat. Within two years the refuge was dotted with special food plots for deer, its occasional trespassers were being apprehended, and an intricate burning plan was designed by fire boss Jim Durrwachter to alternate fires through all of the refuge's uplands and prairies. For the first time since recovery began in 1976, a part of Florida was being managed specifically for the panther. A year-round staff of six plus ten winter-season fire technicians dedicated themselves to the task of creating the panther management field guide.

After five years of the best professional brains, the best equipment, and the best management techniques, panthers responded by continuing to use the refuge exactly as before. Was this another failed experiment? Hardly. Perhaps insufficient time had passed to allow panthers to respond to presumably better management. But it is compelling to think that panthers perceived no difference between the self-imposed restraint of private management (without taxpayer support) and the practices imposed by federal mandate (at a cost of some $500,000 annually, not counting the cost of the land itself). In other words, we observed no differences in panther distribution, abundance, and behavior after the land's acquisition. At the least, Todd Logan should be congratulated for maintaining the legacy that had been handed him—that is, continued panther use in an area bordered by intensive agriculture.

Given the year-round work of refuge management staff, today's level of human activity may well exceed that imposed by its former custodians, who included a handful of cattlemen and a small group of camp owners who managed the property at their own expense. It is very likely that the 25,000 acres that are now the Florida Panther National Wildlife Refuge demonstrate the zone of overlap between tolerable human activity and utterly undisturbed panthers—a broad band of demarcation within which panthers, a variety of management approaches, and human activities are completely compatible with each other.

* * *

What do these unintentional experiments tell us about the manageability of the panther? First, not all south Florida landscape parcels are equally capable of supporting these animals. Panther and panther prey abundance are, and always have been, a function of the land's productivity—that is, soil, drainage, and vegetation patterns are the basic determinants of panther abundance and distribution. Second, the simple elimination of deer hunting, as was done in Everglades National Park and Fakahatchee Strand State Preserve, does not correct other inherent local problems. Panthers do not select home ranges based solely on their tranquility. Third, habitat fragmentation as exhibited by the Golden Gate Estates encourages uncontrolled human access and activities that appear to repel panthers. With appropriate management, though, panthers will flourish in the southern Golden Gate Estates. Fourth, given adequate prey and cover as are found at the Ford Vehicle Evaluation Center, efforts aimed at excluding panthers will fail. Fences may help panthers find underpasses, but they will not prevent them from obtaining their next meal. Fifth, panthers are very tolerant of high levels of human disturbance, such as hunting in Bear Island, so long as these human activities are systematic and predictable. Managers should consider this a very encouraging finding with respect to south Florida's increasing human population and concomitant demands for outdoor recreation. And sixth, the epitome of management that is practiced at the Florida Panther National Wildlife

Refuge is good for panthers, but it is no better than the private management that preceded it. Not only is panther management on private land possible, but it would save millions in taxpayer dollars if landowners were partners in panther recovery.

Hidden between the lines of these observations is a Florida panther conservation plan that can work. The same conclusion was reached after several GIS analyses of south Florida panther habitat in the early 1990s. Jim Cox of the GFC's Office of Environmental Services and I collaborated to identify parcels of unprotected panther habitat. The GFC did not disseminate our findings, however, so the report had little impact. At about the same time, Frank Mazzotti and his associates at the University of Florida analyzed the impact on wildlife of full citrus development in southwest Florida. Focusing on the Florida panther and its habitat, they developed a localized forecast for the South Florida Water Management District. This study was highly regarded by many local farmers, businesspersons, and government agricultural extension workers. The warm reception of this document, which also mapped key parcels of privately owned panther habitat, can be traced to two factors: a clear purpose and the involvement of private property owners from the study's inception through its completion. Even though the report concluded that the full expansion of citrus in south Florida would be disastrous to the panther, no one criticized this document. Unlike the Panther Habitat Preservation Plan, it was not created in a vacuum. Neither the study nor its findings were a surprise to anyone, and its scientific credibility was not compromised. Despite their success, Mazzotti and his colleagues have not been invited to collaborate in agency planning.

Although these landscape analyses seem to have done little to protect panther habitat on private land, the University of Florida study did create a solid foundation for future efforts. There is little question where we need to direct these efforts. Similarly, there is little disagreement that monetary incentives will be essential if we want to enlist landowners as panther conservationists. The biology of the panther has been thoroughly described. Perhaps it is time now to divert research dollars to actual management and incentives on private lands. Since about $500,000 is spent annually to manage the land used by four adult panthers on the 25,000-acre Florida Panther National Wildlife Refuge, the cost of supporting one panther is about $5 an acre. This is considerably less than buying land at $1000 per acre and then adding the perpetual cost of government management. In fact, $5 an acre may be just enough for a landowner to reconsider the future of his or her property. Subtle adjustments to government spending on panther research and management—and tapping into the Florida Panther Trust Fund (a substantial but largely ignored source of funding)—would easily cover the cost of subsidizing property owners in supporting panthers within the habitat core. If a landowner can demonstrate that more than one panther uses the property, the per acre compensation should rise accordingly.

Presumably the recovery of endangered species is for the public good. Thus it is not unreasonable to count on public dollars to make it happen. Owners of land that supports panthers should not be penalized for having panther habitat. Without these lands as an integral part of the habitat core, all the millions of dollars spent on research, public land management, and underpasses will have been wasted. We seem quite willing to spend $5 an acre per panther on a federal refuge. Why shouldn't we be prepared to spend a comparable amount on private lands?

As former Panther Refuge manager Todd Logan has observed, all we need is the money spent for one B-1 bomber to buy all the land needed for panther conservation in south Florida. With the same number of dollars, however, we could leave this land in private ownership, subsidize panther management on it—and probably increase the range of the Florida panther as well. The private sector holds all the keys to panther recovery: its creativity, honed by wrestling with free market problems, will save the panther if we give it the opportunity to do so.

What panthers need is space. The 2 million acres or so still available to them may well be the barest minimum. Ideally, their home should be mostly a forested landscape that can be shared with people. Neither species needs to be excluded. Certainly, panthers will blissfully pursue their day-to-day business in preserves totally devoid of human activity just as they did before humans arrived in Florida thousands of years ago. But they can get along just as well with certain kinds of human activity. Although intensive development would eliminate them, a variety of land management strategies have been shown to be compatible with panthers. The path to panther recovery, therefore, ends on private land.

*   *   *

Panther recovery has been a project of separate agencies exercising their brand of management on separate units of habitat. Unfortunately, none of the agencies represented on the Florida Panther Interagency Committee have the power to effect large-scale changes to the parts of south Florida that contain the best panther habitat. Neither the Florida Park Service nor the U.S. Fish and Wildlife Service controls enough land to make a measurable difference in south Florida's ability to support panthers. The Florida Game and Fresh Water Fish Commission does not own any panther habitat and is therefore little more than an adviser on land management issues. The National Park Service holds title to more land than any other agency, corporation, or individual in south Florida. But because this land is of such little value to the panther—and the agency does not wish to conduct intensive management on behalf of this subspecies—its contribution to recovery will continue to be minimal. And the result of this fragmentation of purpose? No unified vision of the panther's future has been accepted by all the interested parties, public or private.

Because decisions will always be based on the political needs of agencies and the personal agendas of a handful of administrators, the Interagency Committee is more hurdle than help to panther recovery. The National Park Service in particular should be released from further obligations to these efforts. Not only has history demonstrated the NPS's unwillingness to manage single species, but the quality of its land is so poor (except for Bear Island) that there are few changes that could be made to benefit panthers. Conversely, there is little that could be done to make it worse. With designated trails restricting swamp buggy travel in Bear Island and conservative people-management techniques in place, staff of the Big Cypress National Preserve and the GFC should be allowed to argue their opposing philosophies without distracting from landscape-level recovery efforts. National Park Service lands are contributing as much as they can to the population now.

The real potential for progress rests in the hands of the government's water managers: the South Florida Water Management District and the U.S. Army Corps of Engineers. Both agencies have long histories in south Florida. And while their environmental records may contain questionable actions created by outdated management goals, both agencies have the power to make great strides in restoring and protecting much of south Florida. These two organizations have always taken a landscape-scale view of soggy south Florida. Why not form a partnership to advance the cause of panther recovery? This is where the two wildlife agencies could come into their own—acting as the advisers they really are. It is unfair to expect agencies without the land base, or with land too poor to matter, to be responsible for an animal they have virtually no ability to protect. No doubt this is a big reason why technology and animal husbandry have become the primary recovery tools. Captive breeding, genetic manipulation, and other symptom-oriented actions are devices used by agencies unable to do anything else for a space-hungry animal that only needs lots of what the government does not have—good land. And the Army Corps and Water Management District are probably capable of envisioning such landscape repairs as a Caloosahatchee River overpass.

A new private land advisory group should be formed. This group would answer directly to the South Florida Water Management District and the U.S. Army Corps of Engineers. A committee of prominent landowners holding actual or potential panther habitat—in conjunction with field scientists knowledgeable about large carnivore biology and wildlife management—should develop the guidelines that would be implemented by the landscape regulators to benefit panthers living on private land. A prominent member of this group should come from the Big Cypress Seminole Indian Reservation because the Seminoles are, in effect, private owners of an important chunk of panther real estate. And panther conservation planners must listen to their ideas. In many ways, this key group holds the trump cards in the game of panther recovery.

Another critical problem has been research. The politicization of research has slowed progress toward panther recovery and diverted money that could be better spent. Turning over the responsibilities of panther research to an independent academic institution such as the University of Florida's Department of Wildlife Ecology and Conservation could reduce some of the distortions of politics. And research, it should be noted, is not the same as monitoring. Monitoring is the act of keeping an eye on something to see what happens. Research is the act of seeking an answer to an important management question. While agency staff should be encouraged to continue their monitoring activities and participate in developing panther management plans for cooperating landowners, universities are much better equipped to undertake research.

University research, while not immune from politics, operates in an academic setting where experimentation, science, statistics, peer review, and publication are all tenets of existence. Research assistants and graduate students could conduct low-cost fieldwork directed by a select group of faculty who would assure objective science and research continuity. The world-class research on black bears supervised by University of Tennessee professor Mike Pelton through scores of graduate students is a model for such a program. Through a legion of students Pelton has been effective in advising the National Park Service, the U.S. Forest Service, the U.S. Fish and Wildlife Service, and a host of state agencies on black bear issues in the southern Appalachians. Such an arrangement for panthers could even allow natural resource agencies to develop scholarship programs that would eventually bolster their own technical foundations. And, as we have seen in south Florida, university workers may be in a better position to bridge the distrust that private landowners harbor for government agencies.

No matter who does the panther research, the application of the scientific method is a must. University-led research would assure this. In the past, the GFC and the NPS have too easily accepted uncritiqued recommendations on aspects of panther management. This new approach would require long-overdue changes in the middle administrative ranks responsible for these agencies' research. But the value of highly experienced field personnel as an ingredient of research cannot be overemphasized. Thus the Florida Game and Fresh Water Fish Commission should retain responsibility for capturing of individual panthers. As the only agency with a veterinarian on staff for this purpose, and with the most experienced field personnel, the GFC is best suited to carry out all panther captures. This would simplify capture operations, rebuild a team atmosphere, and, ultimately, reduce the likelihood of panther injury or fatality due to inexperience or carelessness.

The U.S. Fish and Wildlife Service, together with the Florida Game and Fresh Water Fish Commission, must acknowledge their strategic error of not involving private landowners from the beginning in the recovery plan. A new recovery plan, with full cooperation from the private sector, is needed. And the importance of maintaining normal demographics should dictate the extent to which south Florida panthers are manipulated in the wild— whether for captive breeding, if it becomes necessary, or for planned genetic introgression if it is proved safe. Because the south Florida population of panthers is viable and functioning normally today, there is time to develop a sane reintroduction program—if progress toward supporting panthers on private lands can be made. Without these private lands, all the well-intended but misplaced efforts of ill-equipped agencies will work like aspirin on a terminal illness. Before real progress can be made, the malaise of government-only panther recovery must be cured.

THE PANTHER'S UNCERTAIN FUTURE

15

Personnel turnover had crippled our capture crew for the start of 1993. Melody Roelke had resigned the previous summer, and so we began the year without a staff veterinarian. Consequently, it was impossible to schedule regular capture activities and impossible, too, to take advantage of an excellent year of panther reproduction by capturing and radio-collaring kittens. With the help of Naples veterinarian John Lanier we muddled through the captures necessary to replace aging transmitters worn by older panthers. Yet even John's recruitment as a volunteer required a plethora of paperwork that further delayed our progress. As it turned out, John donated two or three days each week to help us maintain contact with a study sample that was the product of more than a decade of persistent and dedicated fieldwork.

With each panther handled in the study it became increasingly clear that all of our experiences with these animals could be boiled down to a paragraph of generalizations. The Florida panther is a creature of the landscape. A panther's health and productivity depend on habitat quality, and the area required to support a breeding population crosses many political as well as geographical boundaries. Fragmented forests, flooded soils, and few prey result in ephemeral subpopulations such as those in the Everglades and most of the Big Cypress Swamp. Continuous forest, well-drained soils, and abundant prey support a persistent core population that is tolerant of a variety of human activities. Despite their low numbers, panthers exhibit normal demographics. Genetics have not yet become such a problem for this population that drastic actions with unknown consequences are necessary. Because they are dominated by habitats that panthers tend to avoid, south Florida's public lands alone will not guarantee a future for panthers. Private lands make up a significant portion of the panther habitat core in southwest Florida, and they may totally encompass a second population in south central Florida. Insufficient efforts have been made to assess the panther's status north of the Caloosahatchee River, no effort has been made to relink both sides of the river and too little has been done to enlist the private landowner in panther recovery on the south side of the river. The continued treatment of symptoms will not improve the panther's survival probabilities.

One of my last panther captures was that of Number 42, one of the two refugees from the now-extinct Everglades population. My staff had come to dread captures of panthers in the Big Cypress National Preserve, and Number 42 lived in the heart of it. Past captures of panthers in the preserve were an exercise in frustration because of the conflicting emotions and opinions of National Park Service staff. They preferred consensus over leadership. They would vote to determine if a particular panther was catchable or if there were any reservations about proceeding. I cannot remember a single capture that did not fill me with doubt about its outcome. It was like the dread of poisonous snakes—if you thought about them too much it would ruin your day and make each step through the swamp a fearful struggle. In the Big Cypress National Preserve, the staff's confusion led to excuses for delaying captures that ranged from the presence of large trees to the way a capture had gone the day before.

One time we were forced to attempt a dangerous capture of Number 23, the other Everglades escapee, because of a change in heart. It was a sweltering day in June. Her original capture date, well within the limits of her transmitter's battery life, had been postponed. But now Number 23's radio collar had ceased to emit a regular signal and the situation was urgent. Because the delay went knowingly beyond this limit—as if faith alone would keep the equipment running—a panther's life was put at risk. The combination of summer heat, humidity, and the dogs' treeing the panther in a thick unclimbable strangler fig nearly resulted in her death. Losing her grip on one of the tree's few branches, the drugged panther swung out like an unattached pendulum, fell 30 feet in an arc to the ground, and landed in a motionless lump 6 feet from the edge of the crash bag. At least she had missed the many cypress knees projecting from the ground. Somehow she survived the fall unscathed.

4 May 1993 ■ There was no such pressure to recapture Number 42. We made a long drive on swamp buggies during a hot spring morning to his daytime rest site. Park Service biologist Deb Jansen led us to the vicinity of the cat, winding through an old slash pine forest with trees containing red-cockaded woodpecker cavities and stunted cypress trees adorned with the webs of crab spiders and bromeliads. Roy's son Rowdy was our houndsman during this month of disrupted capture activities. Deb had brought along as field assistants two office workers, a graying midlevel administrator and a GIS operator. Rounding out the complement of novice helpers were Rowdy's youthful hounds and Mike Dunbar, the GFC's new veterinarian who was on his first panther capture. Despite all the newcomers, the experienced hands of Jayde, Rowdy, and Darrell, as well as those of John Lanier, now also a veteran, would be more than enough to succeed.

4 May 1993 ▪ After waiting an hour under a hot morning sun, we left Oasis Ranger Station at 8:30 A.M., bouncing north over pinnacle rock and spinning through mud as the temperature climbed to summerlike heat. Captures during the month of May were unusual. Most often they were reserved for the early morning recaptures of growing kittens in order to expand their collars. About one-quarter of a mile from Number 42, a dead limb, hidden in the mire of a mud hole, was thrust by a rear tire into the drive train of my swamp buggy. With a jolt and a clang the vehicle came to a stop. The clamps securing the universal joint had snapped and the cylindrical bearings were scattered, useless, in the mud. I told Rowdy and Jayde to proceed with the dogs and go after the panther while I tended to my stricken buggy. After a while, using only front-wheel drive, I was able to limp the buggy with its disjointed drive train up the trail and join the others who were now awaiting a signal from Rowdy.

At first Number 42 treed in a short cypress tree. We hurried through sparse palmettos and tussocks of wetland grasses and sedges to inflate the crash bag, but without all available hands contributing (the videocamera had to be held) our progress was slow. The cat, apparently feeling insecure only 20 feet up, leaped while Rowdy tried in vain to home a dart into the suddenly airborne panther. The chase was on again!

Leaving Rowdy and the hounds alone to pursue the panther, we repacked equipment, then slogged deeper into the swamp following the sounds of dogs and the beeping of the radio receiver. The occasional splashing of soggy boots and sneakers on wet prairie grasses quickly turned into the sloshing of legs thigh-deep in an inundated cypress dome. Summer had arrived early this year and the swamps were just about saturated.

This time Number 42 treed 30 feet up in an impressive cypress, flat-topped with age, asymmetrical with wind and lightning damage, and peppered with bromeliads. Although he could not go much higher, he was perched in a strange, nearly vertical position. Rowdy had trouble getting a good view of the panther and five darts missed before one found its mark. Although the syringe was loaded for a 120-pound animal, after fifteen minutes the cat was still too alert to fall or to handle. Apparently only a partial dose had been administered. But we were running out of darts, and could not afford any more misses, so Jayde began a slow, careful ascent with dart pistol in hand. At such close range, he could hardly miss.

But in keeping with the progress of the day, at the moment Jayde squeezed the trigger, the cat shifted, the dart disappeared into a blinding blue sky, and the whoosh of compressed air forced Number 42 back into action. The panther scrambled to reposition himself then lost his grip on slippery bark before falling awkwardly but squarely into the center of the waiting net and crash bag. Ordinarily, as soon as a panther hits the net, the loose ends are thrown over it to prevent escape. But with Jayde in the tree there was too little experience in the water to block the cat's exit. The panther had several pathways open to him. Only one led to dry ground. Within milliseconds he was gone, heading deeper into the swamp.

Suddenly our half-drugged panther was dog-paddling into a band of pop-ash trees that projected out of deep water under the shade of taller cypress. Without a word Rowdy and I bolted after Number 42 while taking turns plunging headfirst into the cold black waters of unseen gator holes. I had never seen a panther swim. If it weren't for the potentially disastrous circumstances, it would have been a marvelous sight. Only his head remained above water while his legs churned below, leaving a wide V-wake in his path—like a gigantic blonde otter. We were gaining on him, but only because the water was so deep his feet could not touch bottom. Still, a nagging question remained: what would we do if we did get our hands on him?

He foundered on a tiny pop-ash island, his shoulders just above water, where it seemed he might rest. Drugs and exhaustion had taken their toll. He paused, panted heavily, but did not continue. Without waiting to make a cautious approach, I groped for his submerged tail, grasped it, and pulled tight expecting his surge in the opposite direction. Within seconds Rowdy appeared and took the tail while I crept closer to Number 42's head. I grabbed his collar and called for a syringe with more Telozol. Rowdy and I restrained the panther until more help arrived, but he twisted and snarled, bringing sharp teeth menacingly close to my exposed right arm. The only recourse was to force the cat's head under water for a moment to change his demeanor. This strategy seemed to work.

After what seemed an eternity—no more than five minutes really—John waded to us to administer the final dose that would keep Number 42 controllable for his workup. Perhaps the only benefit of this pandemonium was the water that kept the panther cool. The rest of us were exhausted and, despite the water, dehydrated. Number 42, like nearly all of the panthers we handled, made an uncomplicated recovery following his ordeal. When he appeared to be alert and peering from a small clump of saw palmetto, we retraced our steps to our morning starting point. Even my jerry-rigged buggy made it home.

*   *   *

Driving west to Naples on U.S. Highway 41, I reflected on the day's events—a capture complicated by harsh environmental conditions and unqualified assistants. The sun glared through rising cumulus as I squinted and wondered. How had we so often escaped disaster when administrators in an office had undermined efficient field operations by turning captures into question-and-answer sessions? A matter of inches determined whether a panther hit the ground or the net, and this took experienced hands. Yet even with experience, many aspects of each capture were beyond our control. Perhaps it was divine intervention that spared me from witnessing a capture tragedy. But I think the real reason that panthers weathered our intrusions so well was their inherent toughness, evolved over thousands of years in one of the harshest environments in North America. This individual ruggedness is a by-product of the species' adaptability. It's a toughness that has enabled the panther to survive assaults on its home as well as our insignificant efforts to help it.

How many more years would go by before the annual flurry of captures and unending radio tracking finally convinced administrators what needed to be done? Few studies of large cats have had the luxury of funding for fifteen consecutive years of fieldwork. We have learned a great deal. The natural history questions have been answered, and the panther's future is now a question of policy. This is why I suggested that natural history research was no longer necessary—that any continued work with panthers should simply be a matter of monitoring to measure their responses, if any, to large-scale management programs. If my suggestions were ignored, perhaps it was because they would erode the funding-based power of research administrators. Today there are many better ways to use the hundreds of thousands of dollars spent each year on panther research. Not that the money was wasted. Clearly, some progress has been made, not only in understanding panther biology, but in effecting modest landscape changes that will benefit many species of wildlife. The money spent on underpasses and land acquisition will pay long-lasting dividends for virtually every species of south Florida wildlife from bobcats to black bears, even if the end of protecting additional panther habitat is at hand. Thus even if panthers disappear, the species that remain, as well as the landscape in which they live, will have better prospects for survival.

By the end of the year, Jayde would be transferred away from the project.

Walt had already been in north Florida for several months. Their departures were all voluntary, but reluctant. I tried everything to promote them within the panther project in order to maintain the irreplaceable knowledge they brought to panther recovery. Unfortunately, leaving was the only way they could obtain the compensation they deserved and escape a situation where they felt unappreciated.

Watching the sun disappear behind a growing thunderhead, my mind wandered to the panther's real problems: too many people spread randomly across the landscape and too many people still coming to south Florida. U.S. 41 was now whisking me past sprawling mobile home parks, their owners dependent on drainage canals, air conditioning, and mosquito control for a habitable environment. The technology of instant housing had spurred the expansion of urban boundaries beyond the sleepy fishing and resort communities that once hugged the coastline. These inland blights of human population sprawl were recent. Many had appeared before our eyes like lawn mushrooms that spring up over a summer night. Like all animal species, including panthers, humans not only have an irrepressible drive to procreate and expand their numbers, but also to increase the limits of their distribution. While these traits were once keys to our survival, they now threaten not only a wide array of plant and animal species but entire ecosystems as well. Given unlimited space, panthers too could increase without artificial bounds.

Summer evenings in south Florida are often spectacular. The daily afternoon thunderstorms create the billowing cloud formations that relieve this region's flatness, and their electricity punctuates the clouds' colorful illusion of topography. Thunderstorms are the most visible evidence of the vitality and power of ecological processes that have, for millennia, molded the south Florida landscape and the creatures that live here. They speak of the naturalness that remains—events that are witnessed equally well from a fashionable clothing store on Fifth Avenue in Naples as from the edge of a tropical hammock at the center of east Hinson Marsh in Bear Island. The storms are the product of forces that bind together these alien worlds: the domain of Florida's most widespread mammal, and the haunts of its rarest.

As I approached the city limits I could imagine afternoon beachgoers scurrying for shelter from picnic-ending lightning and rain. I could also visualize panthers tucked away in saw palmetto shelters. The rain would bring coolness and signal the stirrings of prey. I could imagine an adult female panther emerging silently from dripping palmetto fronds and squinting briefly at the pink, flashing thunderheads as they retreated to the east and evaporated into darkness. Behind her, a pile of well-fed kittens lay clumped and silent amid the tangled roots of palmetto and twisting vines of poison ivy. Their mother would faithfully return after her hunt. Although she would now be hungry after a full twenty-four hours of nursing, grooming, and tolerating the play of her growing brood, she is well-muscled, sleek, and in perfect condition to maintain herself and her kittens. Finding the next meal would not be a problem. Perhaps her kittens would be raised, undisturbed, by the hounds and people that had captured some of her kin and they would become important cogs in the machinery of panther society. A burst of static from the radio interrupted the daydream.

*   *   *

Remarkably, south Florida still retains much of its primordial landscapes. Because of this essential wildness, the effects of human intrusion have been dampened and, to a large extent, ignored by its remnant population of panthers. Without a new conservation paradigm, however, this resiliency will not last. Continued contraction of the panther's range may not cause increased mortality, but those panthers that continue to succumb to highway collisions and other unnatural events will increasingly represent key members of the population. Instead of losing expendable, dispersing subadults, resident breeders will die and not be replaced. When this begins, home ranges will become vacant, kittens will be lost, and reproduction will finally cease.

There are many who claim that panthers make Florida more special than all the alligators, bromeliads, and tree snails in the forest. Yet there are others equally sure that the landscape has no need for a top-order predator—prey species, they say, will be regulated efficiently enough by bears, floods, bobcats, and their own self-limiting mechanisms. Perhaps it is wrong to think that we need the panther for any practical purpose or to maintain some "ecological balance." And yet it seems criminal that we could let Florida's top carnivore slip into extinction when it may be the most important evolutionary force on white-tailed deer and other prey in south Florida. This force has been lost everywhere else in eastern North America, a region where wildlife communities seem more a part of the human landscape than products of evolution.

Another reason we need the panther is for ourselves. Like the muskrats studied for so many decades by Paul Errington in midwestern marshes, panthers have the power to tell us a lot about ourselves. Not so much because there is anything inherently humanlike about them, but because we share ancient roots in evolution and because they have been studied so long. What has emerged from long-term research such as Errington's and our panther work are behavior patterns analogous to human behavior. The similarities are compelling. When panthers perceived overcrowding, violent deaths resulted. These deaths helped to maintain a balance in the population, and we came to expect these losses as a normal part of social dynamics. Extreme violence, drug abuse, and social decay are equally indicative of overcrowded urban centers where humans appear to have exceeded the carrying capacity of metropolitan infrastructure. Aggressive encounters become more likely as individuals are forced to react to ever-increasing competitors. Even government agencies exhibit a form of territoriality: the fittest do not always survive, but the tenured have prior rights to power maintained by high administrative positions. Just like panthers, the rise of an administrator has more to do with timing and random events than talent; it is a matter of being in the right place at the right time. As the size of cities and bureaucracies grows, so too does the strife within them. No doubt the outbursts of human violence and bureaucratic territorialism are as evolutionarily predictable as the violence we documented among panthers.

Male panthers were the wandering, promiscuous, domineering, and intolerant members of the population. Female panthers were its building blocks and the anchors of social stability. And yet they were the cause of fighting, posturing, and death among males—more so than any other factor, natural or unnatural. It is impossible, for me at least, to ignore the similarities between panthers and the human behavioral stereotypes that Americans so openly abhor but seem so helpless to overcome. The biggest difference between us and panthers is that we can recognize these patterns, accept our biologically programmed character flaws, and do something about them. Perhaps the greatest value in studying endangered species is what we learn about ourselves in our haphazard efforts to save them.

If we allow a self-sustaining population of panthers to exist in Florida, there may be some reason to think we can learn to manage ourselves as well. Instead of gauging human prosperity by our ability to sustain economic growth, we will learn the wisdom of a sustainable, dynamic balance as exhibited by Florida panthers. If nothing else, the panthers we studied demonstrate that coexistence with people is possible. A lasting balance can be built on the efforts of agencies and people who understand that creativity, honesty, and imagination are the keys to success. With changing times and changing paradigms, there is no need yet to write the panther's epitaph.

EPILOGUE

I can think of few examples that illustrate the tenacity of the Florida panther as well as Female 9. Captured in 1985, she still wears a radio collar and continues to inhabit the foreboding Fakahatchee Strand. She has weathered at least six captures, a shotgun blast to the foot, a short bout in captivity to repair broken foot bones, the removal of her last successful litter, and three south Florida panther project leaders. I would guess that she is now about fifteen years old and well past her reproductive prime. Like the first living panther that I saw in 1986, Female 9 has become the grizzled shell that Female 8 once was.

Many other players in panther society have redeemed their tenure on a chunk of south Florida wilderness. Males 12 and 26, both beyond their physical prime, met violent deaths at the claws of competitors and have been replaced by upstart Males 45, 46, and 54. These young males are now the kingpins of the panther habitat core. Female 31 was run over on a desolate Collier County road in early 1994 as her health declined and her newest female offspring clamored for more space. Females 11 and 40 both lost their radio collars but were recaptured a few months later—Female 11 is past her reproductive years, but Female 40 continues to produce kittens. Male 42, the killer of dispersing Male 44, died in the summer of 1995 of unknown causes leaving the eastern Big Cypress/Everglades without a male once again. Seventeen radio collars still transmit the locations of their feline wearers. As much as things have changed, they are still much the same.

But many of the misconceptions also remain, and even the U.S. Postal Service has become a Florida panther advocate. Their two-page ad in the November–December 1996 *Audubon* laments: "Less than 60 left in the Everglades. We've honored them with a stamp." Little did they know that fewer than two were left in this most desolate of panther landscapes. The remaining 98 percent continue their existence, as always, in the more productive lands to the north and west of the River of Grass. But the number of big cats has increased dramatically in the last year. With the introduction of eight female Texas cougars into south Florida, and their production of at least four litters of hybrid kittens, the new wave of panther recovery is under way. Genetically fortified kittens will soon be loosed upon a south Florida landscape that has repeatedly demonstrated its inability to nurture them. It is anyone's guess where this will all lead. There is no blueprint for the builders of the new panther in Florida to know when the work is done.

Early in 1997, males 201 and 203, two of the kittens we removed for captive breeding in 1991, were released in southeast Florida to breed with two of the recently introduced Texas females. It was presumed by panther managers that they were roaming the eastern Big Cypress Swamp without mates. Both wild-born but captive-raised male panthers were dead by the end of March 1997. An official cause of death had yet to be attributed at the time of this writing, but the 40 pounds lost by each of them suggest that malnutrition—exacerbated by low deer numbers and poor hunting skills—may have had something to do with it. Of equal interest is the appearance of an uncollared male in the area of 201's and 203's release—suggesting that the loss of these two captive-raised panthers was unnecessary. Just as a long line of males has inhabited the Fakahatchee Strand, home range vacancies such as these do not stay unoccupied for long. A little patience is all that is needed. Regardless of the results of this most recent experiment, genetically fortified kittens will soon be loosed upon a southeastern Florida landscape that has repeatedly demonstrated its inability to nurture them. But our panthers remain—living, dying, replenishing themselves—just waiting for their chance to show us how well they can respond to expanding landscapes and changing times.

VITAL STATISTICS OF FLORIDA PANTHERS

TABLE A.I.

Florida Panthers Captured in the Wild in South Florida: 1981–1994

ID No.	Sex	First capture weight (pounds)	First capture age	Date of first capture	Date of death[a]	Cause of death[b]	Number of radio locations[c]	Parents if known
01	M	120	10	2/10/81	12/14/83	HBC	47	
02	M	108	10	2/20/81	11/29/84	IA	473	
03	F	68	9	1/23/82	1/17/83	OD	192	
04	M	113	8	1/27/82	4/18/85	HBC	325	
05	F	98	8	2/23/82	11/23/82	?	307	
06	M	122	7	2/27/82	4/16/82	?	0	
07	M	122	6	3/2/82	10/26/85	HBC	231	
08[d]	F	75	7	3/25/84	8/20/88	old age	245	
09	F	79	3	1/26/85	—		1426	
10	M	34	0.4	1/15/86	1/27/87	IA	207	7 & 9
11	F	92	4	1/21/86	—		1412	
12	M	122	5	1/28/86	11/9/94	IA	1406	
13	M	126	4	2/27/86	12/14/87	HBC	313	
14	F	71	5	12/7/86	6/20/91	?	1282	
15	F	72	5	12/13/86	6/10/88	?	505	
16	M	86	1	1/12/87	—		1920	14 & ?
17	M	142	7	1/20/87	7/20/90	?	554	
18	F	100	8	1/22/87	10/1/90	IA	595	
19	F	49	0.7	2/9/87	—		1200	11 & 12
20	M	148	3	3/10/87	8/24/88	HF	215	
21[d]	F	56	1	3/16/87	—		469	14 & ?
22	F	32	0.5	3/18/87	7/22/91	Infection	700	15 & ?
23	F	31	0.5	3/18/87	—		1492	15 & ?
24	M	126	3	1/30/88	8/28/88	?	59	
25	M	121	4	2/16/88	8/26/88	IA	87	
26	M	120	5	3/1/88	7/8/94	IA	954	
27	F	50	2	4/11/88	7/23/89	?	451	
28	M	105	1.5	11/29/88	9/25/92	IA	532	
29	M	44.5	0.5	1/3/89	5/27/92	PR	522	11 & 20
30	M	49	0.7	1/6/89	1/29/90	IA	203	19 & 13
31	F	85	8	1/12/89	3/3/94	HBC	805	
32	F	71	2	2/3/89	—		906	
33	M	92	1.5	3/5/89	11/25/89	Rabies	256	
34	M	62	0.8	1/8/90	11/15/93	Puncture	576	31 & ?

continues

ID No.	Sex	First capture weight (pounds)	First capture age	Date of first capture	Date of death[a]	Cause of death[b]	Number of radio locations[c]	Parents if known
35	M	52	0.8	1/15/90	1/24/90	Infection	0	31 & ?
36	F	108	5	1/27/90	—		723	
37	M	102	4	1/30/90	11/26/90	HBC	128	
38	F	98	4	2/8/90	8/4/94	Infection	927	
39	M	102	3	2/19/90	6/18/90	Infection	81	
40	F	65	2	2/26/90	—		585	
41	F	61	2	2/28/90	9/21/90	IA	89	
42	M	60	0.9	3/6/90	6/22/95	?	766	14 & 16
43	M	54	0.8	5/1/90	11/1/91	IA	222	19 & 12
44	M	33	0.5	4/30/91	7/6/93	IA	272	40 & 26
45	M	37	0.5	5/8/91	—		482	19 & 12
46	M	116	2	1/30/92	—		419	
47	M	42	0.5	2/21/92	2/19/93	IA	150	11 & 12
48	F	25	0.3	2/24/92	—		416	31 & 12
49	F	71	2	2/25/92	?		78	
50	M	62	0.6	3/4/92	12/6/93	HBC	256	36 & 26
51	M	108	3	3/26/92	—		407	
52	F	39	0.5	5/5/92	1/14/95	HBC	387	31 & 12
53	M	60	0.8	2/10/93	2/26/93	IA	7	19 & 12
54	M	66	0.8	2/10/93	—		275	40 & ?
K1[e]	M	2.6	0.05	4/7/92	—		—	40 & ?
K2[e]	M	2.4	0.05	4/7/92	—		—	40 & ?
K3[e]	M	4.0	0.06	6/18/93	—		—	40 & 26
K4[e]	F	4.0	0.06	6/18/93	—		—	40 & 26
K5[e]	F	3.7	0.06	6/18/93	—		—	40 & 26
K6[e]	M	1.3	0.02	10/30/93	—		—	48 & 26
K7[e]	F	1.3	0.02	10/30/93	—		—	48 & 26
K8[e]	F	1.3	0.02	10/30/93	—		—	48 & 26
201[d]	M	34	0.6	2/20/91	3/11/97	PS[f]	—	31 & 12
202[d]	M	47	0.7	2/22/91	—		—	9 & 37
203[d]	M	45	0.7	2/25/91	2/26/97	PS[f]	—	9 & 37
204[d]	F	37	0.6	2/27/91	—		—	31 & 12
205[d]	F	31	0.5	5/3/91	—		—	19 & 12
206[d]	F	33	0.5	5/6/91	—		—	40 & 26
207[d]	M	59	0.7	3/4/92	—		—	36 & 26
208[d]	F	1.3	0.02	6/6/92	—		—	32 & 12
209[d]	F	1.8	0.02	8/20/92	—		—	42 & 23
210[d]	M	1.8	0.02	8/20/92	—		—	42 & 23

[a] No entry for date of death signifies that the study animal is still alive, the transmitter failed, or the study animal was not instrumented (neonate kittens and kittens removed for captive breeding).

[b] HBC = hit by car; IA = intraspecific aggression; OD = drug overdose; HF = congestive heart failure; PR= pseudorabies; puncture = esophageal puncture; ? = unknown.

[c] Total radio locations = 31,936.

[d] Removed from wild.

[e] Neonates handled at natal dens.

[f] PS = probable starvation

BIBLIOGRAPHY

Ackerman, B. B., F. G. Lindzey, and T. P. Hemker. 1984. Cougar food habits in southern Utah. *J. Wildl. Manage.* 48:147–155.

Allen, R. 1950. Notes on the Florida panther, *Felis concolor coryi* Bangs. *J. Mammal.* 31:279–280.

Alvarez, K. 1993. *Twilight of the panther.* Sarasota, Fla.: Myakka River Publishing.

Anderson, A. E. 1983. A critical review of literature on puma (*Felis concolor*). Spec. Rept. 54. Ft. Collins: Colorado Division of Wildlife.

Anderson, A. E., D. C. Bowden, and D. M. Kattner. 1992. The puma on Uncompahgre Plateau, Colorado. Tech. Publ. 40. Ft. Collins: Colorado Division of Wildlife.

Ashman, D., G. C. Christensen, M. L. Hess, G. K. Tsukamoto, and M. S. Wickersham. 1983. The mountain lion in Nevada. P. R. Proj. W-48-15 Final Rep. Nevada Fish and Game Dept.

Bangs, O. 1898. The land mammals of peninsular Florida and the coast region of Georgia. *Proc. Boston Soc. Nat. Hist.* 28:157–235.

Barrone, M. A., M. E. Roelke, J. Howard, J. L. Brown, A. E. Anderson, and D. E. Wildt. 1994. Reproductive characteristics of male Florida panthers: Comparative studies from Florida, Texas, Colorado, Latin America, and North American zoos. *J. Mammal.* 75:150–162.

Bass, O. L., and D. S. Maehr. 1991. Do recent panther deaths in Everglades National Park suggest an ephemeral population? *Nat. Geogr. Res. and Explor.* 7:427.

Beck, T.D.I. 1991. Black bears of west-central Colorado. Tech. Publ. 39. Ft. Collins: Colorado Division of Wildlife.

Beier, P., and S. Loe. 1992. A checklist for evaluating impacts to wildlife movement corridors. *Wildl. Soc. Bull.* 20:434–440.

Beier, P., D. Choate, and R. H. Barrett. 1995. Movement patterns of mountain lions during different behaviors. *J. Mammal.* 76:1056–1070.

Bekoff, M. 1982. Coyote. In J. A. Chapman and G. A. Feldhammer, eds., *Wild mammals of North America.* Baltimore: Johns Hopkins University Press.

Belden, R. C., and D. J. Forrester. 1980. A specimen of *Felis concolor coryi* in Florida. *J. Mammal.* 61:160–161.

Belden, R. C., and B. W. Hagedorn. 1993. Feasibility of translocating panthers into northern Florida. *J. Wildl. Manage.* 57:388–397.

Belden, R. C., W. B. Frankenberger, R. T. McBride, and S. T. Schwikert. 1988. Panther habitat use in southern Florida. *J. Wildl. Manage.* 52:660–663.

Bertram, B.C.R. 1979. Serengeti predators and their social systems. In A.R.E. Sinclair and M. Norton-Griffiths, eds., *Serengeti: Dynamics of an eco-system.* Chicago: University of Chicago Press.

Brady, J. R., and H. W. Campell. 1983. Distribution of coyotes in Florida. *Fla. Field Nat.* 11:40–41.

Brenneman, R. L., and S. M. Bates. 1984. *Land-saving action.* Covelo, Calif.: Island Press. 249pp.

Burghard, A. 1969. *Alligator Alley: Florida's most controversial highway.* Washington, D.C.: Lanman.

Burt, W. H. 1975. *A field guide to the mammals.* 3rd ed. Boston: Houghton Mifflin.

Captive Breeding Specialist Group. 1992. Genetic management strategies and population viability of the Florida panther. Report of a workshop. Yulee, Fla.: White Oak Conservation Center. 27pp.

Carter, L. J. 1974. *The Florida experience: Land and water policy in a growth state.* Baltimore: Johns Hopkins University Press.

Case, T. J., and M. E. Gilpin. 1974. Interference competition and niche theory. *Proc. Nat. Acad. Sci. USA* 71:3073–3077.

Caughley, G. 1994. Directions in conservation biology. *J. Animal Ecol.* 63:215–244.

Caughley, G., and A.R.E. Sinclair. 1994. *Wildlife ecology and management.* Boston: Blackwell.

Chapman, J. A., and G. A. Feldhammer. 1982. *Wild mammals of North America.* Baltimore: Johns Hopkins University Press.

Chepko-Sade, B. D., W. M. Shields, J. Berger, Z. T. Halpin, W. T. Jones, L. L. Rogers, J. P. Rood, and A. T. Smith. 1987. The effects of dispersal and social structure on effective population size. In B. D. Chepko-Sade and Z. T. Halpin, eds., *Mammalian dispersal patterns: The effects of social structure on population genetics.* Chicago: University of Chicago Press.

Cory, C. B. 1896. *Hunting and fishing in Florida.* 2nd ed. Boston: Estes & Lauriat.

Cox, J., R. Kautz, M. MacLaughlin, and T. Gilbert. 1994. Closing the gaps in Florida's wildlife habitat conservation system. Tallahassee: Florida Game and Fresh Water Fish Commission.

Craighead, F. C. 1971. *The trees of south Florida.* Vol. 1. Coral Gables: University of Miami Press.

Craighead, J. J., J. S. Sumner, and J. A. Mitchell. 1995. *The grizzly bears of Yellowstone: Their ecology in the Yellowstone ecosystem, 1959–1992.* Washington, D.C.: Island Press.

Darwin, C. 1871. *The descent of man, and selection in relation to sex.* London: J. Murray.

Davis, J. H., Jr. 1943. The natural features of southern Florida, especially the vegetation and the Everglades. *Fla. Geol. Surv. Bull.* 25. Tallahassee.

De Bellevue, E. B. 1976. Hendry County: An agricultural district in a wetland region. South Florida Study. Gainesville: Center for Wetlands, University of Florida.

Diamond, J. 1992. *The third chimpanzee.* New York: HarperCollins.

Dibello, F. J., S. M. Arthur, and W. B. Krohn. 1990. Food habits of sympatric coyotes, *Canis latrans,* red foxes, *Vulpes vulpes,* and bobcats, *Lynx rufus,* in Maine. *Canadian Field-Nat.* 104:403–408.

Dixon, K. R. 1982. Mountain lion (*Felis concolor*). In J. A. Chapman and G. A. Feldhammer, eds., *Wild mammals of North America.* Baltimore: Johns Hopkins University Press.

Douglas, M. S. 1988. *The Everglades river of grass.* Rev. ed. Sarasota, Fla.: Pineapple Press.

Duever, M. J., J. E. Carlson, J. F. Meeder, L. C. Duever, L. H. Gunderson, L. A. Riopelle, T. R. Alexander, R. L. Myers, and D. P. Spangler. 1986. *The Big Cypress National Preserve.* Research Report 8. New York: National Audubon Society.

Eisenberg, J. F. 1989. *Mammals of the neotropics: The northern tropics.* Vol. 1. Chicago: University of Chicago Press.

Elliott, D. G. 1901. A list of mammals obtained by Thaddeus Surber in North and South Carolina, Georgia, and Florida. *Field Col. Mus. Zool. Series* 3:31–57.

Eloff, F. C. 1973. Lion predation in the Kalahari Gemsbok National Park. *J. S. Afr. Wildl. Manage. Assoc.* 3:59–63.

Errington, P. L. 1957a. *Of men and marshes.* Ames: Iowa State University Press.

———. 1957b. Of population cycles and unknowns. *Cold Spring Harbor Symp. on Quant. Biol.* 22:287–300.

Ewel, J. J. 1990. Introduction. In R. L. Myers and J. J. Ewel, eds., *Ecosystems of Florida.* Orlando: University of Central Florida Press.

Ewel, J. J., D. S. Ojima, D. A. Karl, and W. F. DeBusk. 1982. Schinus in successional ecosystems of Everglades National Park. Report T-676. Homestead, Fla.: National Park Service.

Ewer, R. F. 1968. *Ethology of mammals.* London: Elek Science.

Fernald, E. A., and E. D. Purdum. 1992. *Atlas of Florida.* Gainesville: University Press of Florida.

Forrester, D. J. 1992. Parasites and diseases of wild mammals in Florida. Gainesville: University Press of Florida.

Forrester, D. J., J. A. Conti, and R. C. Belden. 1985. Parasites of the Florida panther (*Felis concolor coryi*). *Proc. Helminthol. Soc. Wash.* 52:95–97.

Foster, M. L. 1992. Effectiveness of wildlife crossings in reducing animal/auto collisions on Interstate 75, Big Cypress Swamp, Florida. M.S. thesis, University of Florida, Gainesville.

Glass, C. M., R. G. McLean, J. B. Katz, D. S. Maehr, C. B. Cropp, L. J. Kirk, A. J. McKeirnan, and J. F. Evermann. 1994. Isolation of pseudorabies (Aujeszky's disease) virus from a Florida panther. *J. Wildl. Dis.* 30: 180–184.

Greiner, E. C., M. E. Roelke, C. T. Atkinson, J. P. Dubey, and S. D. Wright. 1989. *Sacrosystis* spp. in muscles of free-ranging Florida panthers and cougars (*Felis concolor*). *J. Wildl. Dis.* 25:623–628.

Guenther, D. D. 1980. Home range, social organization, and movement patterns of the bobcat *Lynx rufus,* from spring to fall in south-central Florida. M.A. thesis, University South Florida, Tampa.

Harlow, R. F. 1959. An evaluation of white-tailed deer habitat in Florida. Tech. Bull. 5. Tallahassee: Florida Game and Fresh Water Fish Commission.

Harris, L. D. 1984. *The fragmented forest.* Chicago: University of Chicago Press.

———. 1988. The nature of cumulative impacts on biotic diversity of wetland vertebrates. *Environ. Manage.* 12:675–693.

———. 1990. What does geology have to do with panther conservation? *Nat. Geographic Res. and Explor.* 7:117–119.

Harris, L. D., and W. Cropper. 1992. Between the devil and the deep blue sea: Implications of climate change for Florida's flora and fauna. In R. Peters and T. Lovejoy, eds., *Global warming and biological diversity.* New Haven: Yale University Press.

Harris, L. D., and P. Gallagher. 1989. New initiatives for wildlife conservation, the need for movement corridors. In G. Mackintosh, ed., *In defense of wildlife: Preserving communities and corridors.* Washington, D.C.: Defenders of Wildlife.

Harris, L. D., and J. G. Gosselink. 1990. Cumulative impacts of bottomland hardwood forest conversion on hydrology, water quality, and terrestrial wildlife. In J. G. Gosselink, L. C. Lee, and T. A. Muir, eds., *Ecological processes and cumulative impacts: Illustrated by bottomland hardwood wetland ecosystems.* Chelsea, Mich: Lewis.

Harris, L. D., T. Hoctor, D. Maehr, and J. Sanderson. 1996. The role of networks and corridors in enhancing the value and protection of parks and equivalent areas. Pages 173–197 in R. G. Wright, ed., *National parks and protected areas: Their role in environmental protection.* Cambridge, Mass.: Blackwell Science.

Harrison, D. J. 1992. Dispersal characteristics of juvenile coyotes in Maine. *J. Wildl. Manage.* 56:128–138.

Hedrick, P. W. 1995. Gene flow and genetic restoration: The Florida panther as a case study. *Conserv. Biol.* 9:996–1007.

Hela, I. 1952. Remarks on the climate of southern Florida. *Bull. Marine Sci. of Gulf and Caribbean* 2:438–447.

Hemker, T. P., F. G. Lindzey, and B. B. Ackerman. 1984. Population characteristics and movement patterns of cougars in southern Utah. *J. Wildl. Manage.* 48:1275–1284.

Henry, J. A., K. M. Portier, and J. Coyne. 1994. *The climate and weather of Florida.* Sarasota, Fla.: Pineapple Press.

Hibben, F. C. 1995. *Hunting American lions.* Reprint. Silver City, N.M.: High-Lonesome Books.

Hopkins, R. A. 1989. Ecology of the puma in the Diablo Range, California. Ph.D. dissertation, University of California, Berkeley.

Horn, H. S. 1983. Some theories about dispersal. In I. R. Swingland and P. J. Greenwood, eds., *The ecology of animal movements.* Oxford: Clarendon Press.

Hornocker, M. G. 1969. Winter territoriality in mountain lions. *J. Wildl. Manage.* 33:457–464.

———. 1970. An analysis of mountain lion predation upon mule deer and elk in the Idaho Primitive Area. *Wildl. Monogr.* 21:1–39.

Hunter, M. D., and P. W. Price. 1992. Natural variability in plants and animals. In M. D. Hunter, T. Ohgushi, and P. W. Price, eds., *Effects of resource distribution on animal-plant interactions.* San Diego: Academic Press.

Hutchinson, G. E. 1957. Concluding remarks. *Cold Spring Harbor Symp. on Quant. Biol.* 22:415–427.

Jalkotzy, M., I. Ross, and J. R. Gunson. 1992. Management plan for cougars in Alberta. Edmonton: Alberta Forestry, Lands, and Wildlife.

Johnson, M. K., R. C. Belden, and D. R. Aldred. 1984. Differentiating mountain lion and bobcat scats. *J. Wildl. Manage.* 48:239–244.

Jordan, D. B. 1993. Preliminary analysis of potential Florida panther reintroduction sites. Gainesville, Fla.: U.S. Fish and Wildlife Service.

Kautz, R. S., D. T. Gilbert, and G. M. Mauldin. 1993. Vegetative cover in Florida based on 1985–1989 Landsat Thematic Mapper imagery. *Florida Scientist* 56:135–154.

Koehler, G. M., and M. G. Hornocker. 1991. Seasonal resource use among mountain lions, bobcats, and coyotes. *J. Mammal.* 72:391–396.

Koford, C. B. 1946. A California mountain lion observed stalking. *J. Mammal.* 27:274–275.

Laing, S. P., and F. G. Lindzey. 1993. Patterns of replacement of resident cougars in southern Utah. *J. Mammal.* 74:1056–1058.

Land, E. D., D. S. Maehr, J. C. Roof, and J. W. McCown. 1993. Mortality patterns of female white-tailed deer in southwest Florida. *Proc. Annu. Conf. Southeast. Fish and Wildl. Agencies* 47:176–184.

Lande, R. 1988. Genetics and demography in biological conservation. *Science* 241:1455–1460.

Layne, J. N., and M. N. McCauley. 1976. Biological overview of the Florida panther. In P.C.H. Pritchard, ed., *Proceedings of the Florida panther conference.* Casselberry: Florida Audubon Society.

Layne, J. N., and D. A. Wassmer. 1988. Records of the panther in Highlands County, Florida. *Florida Field Nat.* 16:70–72.

Lehner, P. N. 1976. Coyote behavior: Implications for management. *Wildl. Soc. Bull.* 4:120–126.

Leighty, R. G., M. B. Marco, G. A. Swenson, R. E. Caldwell, J. R. Henderson, O. C. Olson, and G. C. Wilson. 1954. Soil survey of Collier County, Florida. Gainesville: Florida Agricultural Experiment Station, University of Florida.

Leopold, A. 1933. *Game management.* New York: Charles Scribners.

Logan, K. A., and L. L. Irwin. 1985. Mountain lion habitats in the Big Horn Mountains, Wyoming. *Wildl. Soc. Bull.* 13:257–262.

Logan, K. A., L. L. Irwin, and R. R. Skinner. 1986. Characteristics of a hunted mountain lion population in Wyoming. *J. Wildl. Manage.* 50:648–654.

Logan, K. A., L. L. Sweanor, J. F. Smith, B. R. Spreadbury, and M. G. Hornocker. 1990. Ecology of an unexploited mountain lion population in a desert environment. Annual report. Moscow, Idaho: Wildlife Research Institute.

Logan, T., A. C. Eller, Jr., R. Morrell, D. Ruffner, and J. Sewell. 1994. Florida panther habitat preservation plan. Gainesville: Florida Panther Interagency Committee.

Maehr, D. S. 1990. The Florida panther and private lands. *Conserv. Biol.* 4:167–170.

———. 1992. Florida panther. In S. R. Humphrey, ed., *Rare and endangered biota of Florida.* Vol. 1: *Mammals.* Gainesville: University Press of Florida.

———. 1996. The comparative ecology of bobcat, black bear, and Florida panther in south Florida. Ph.D. dissertation, University of Florida, Gainesville.

Maehr, D. S., and G. B. Caddick. 1995. Demographics and genetic introgression in the Florida panther. *Conserv. Biol.* 9:1295–1298.

Maehr, D. S., and J. A. Cox. 1995. Landscape features and panthers in Florida. *Conserv. Biol.* 9:1008–1019.

Maehr, D. S., and J. N. Layne. 1996. The saw palmetto. *Gulfshore Life* 26(7):38–51, 56–58.

Maehr, D. S., and C. T. Moore. 1992. Models of mass growth for 3 North American cougar populations. *J. Wildl. Manage.* 56:700–707.

Maehr, D. S., E. D. Land, J. C. Roof, and J. W. McCown. 1989a. Early maternal behavior in the Florida panther (*Felis concolor coryi*). *Amer. Midl. Nat.* 122:34–43.

———. 1989b. First reproduction of a panther (*Felis concolor coryi*) in southwestern Florida, U.S.A. *Mammalia* 53:129–131.

———. 1990. Day beds, natal dens, and activity of Florida panthers. *Proc. Annu. Conf. Southeast. Fish and Wildl. Agencies* 44:310–318.

Maehr, D. S., J. C. Roof, E. D. Land, J. W. McCown, R. C. Belden, and W. B. Frankenberger. 1989c. Fates of wild hogs released into occupied Florida panther home ranges. *Florida Field Nat.* 17:42–43.

Maehr, D. S., R. C. Belden, E. D. Land, and L. Wilkins. 1990. Food habits of panthers in southwest Florida. *J. Wildl. Manage.* 54:420–423.

Maehr, D. S., E. D. Land, and J. C. Roof. 1991a. Social ecology of Florida panthers. *Nat. Geogr. Res. and Explor.* 7:414–431.

Maehr, D. S., E. D. Land, and M. E. Roelke. 1991b. Mortality patterns of panthers in southwest Florida. *Proc. Annu. Conf. Southeast. Fish and Wildl. Agencies* 45:201–207.

Maehr, D. S., J. C. Roof, E. D. Land, J. W. McCown, and R. T. McBride. 1992. Home range characteristics of a panther in south central Florida. *Florida Field Nat.* 20:97–103.

Maehr, D. S., E. C. Greiner, J. E. Lanier, and D. Murphy. 1995. Notoedric mange in the Florida panther (*Felis concolor coryi*). *J. Wildl. Dis.* 31: 251–254.

Mayr, E. 1976. *Evolution and the diversity of life.* Cambridge: Belknap Press.

McArthur, R. H. 1972. *Geographical ecology.* Princeton: Princeton University Press.

McBride, R. T. 1976. The status and ecology of the mountain lion *Felis concolor stanleyana,* of the Texas–Mexico border. M.S. thesis, Sul Ross State University, Alpine, Texas.

McBride, R. T., R. M. McBride, J. L. Cashman, and D. S. Maehr. 1993. Do mountain lions exist in Arkansas? *Proc. Annu. Conf. Southeast. Fish and Wildl. Agencies* 47:394–402.

McCown, J. W., D. S. Maehr, and J. Roboski. 1990. A portable cushion as a wildlife capture aid. *Wildl. Soc. Bull.* 18:34–36.

McCown, J. W., M. E. Roelke, D. J. Forrester, C. T. Moore, and J. C. Roboski.

1991. Physiological evaluation of two white-tailed deer herds in south-ern Florida. *Proc. Annu. Conf. Southeast. Assoc. Fish and Wildl. Agencies* 45:81–90.

McIver, S. B. 1994. *Dreamers, schemers and scalawags.* Sarasota, Fla.: Pineapple Press.

McIvor, D. E., J. A. Bissonette, and G. S. Drew. 1995. Taxonomic and con-servation status of the Yuma mountain lion. *Conserv. Biol.* 9: 1033–1040.

McLaughlin, G. S., M. Orstbaum, D. J. Forrester, M. E. Roelke, and J. R. Brady. 1993. Hookworms of bobcats (*Felis rufus*) from Florida. *J. Helminthol. Soc. Wash.* 60:10–13.

McPherson, B. F. 1984. The Big Cypress Swamp. In P. J. Gleason, ed., *Envi-ronments of south Florida: Present and past.* Vol. 2. Coral Gables: Miami Geological Society.

Mech, L. D. 1983. *Handbook of animal radio-tracking.* Minneapolis: Univer-sity of Minnesota Press.

Mech, L. D. 1987. Age, season, distance, direction, and social aspects of wolf dispersal from a Minnesota pack. Pages 55–74 in B. D. Chepko-Sade and Z. T. Halpin, eds., *Mammalian dispersal patterns: The effects of social structure on population genetics.* Chicago: University of Chicago Press.

Meffe, G. K., and C. R. Carroll. 1994. *Principles of conservation biology.* Sun-derland, Mass.: Sinauer Associates.

Mehrer, C. F. 1975. Some aspects of reproduction in captive mountain lions *Felis concolor,* bobcat *Lynx rufus,* and lynx *Lynx canadensis.* Ph.D. dis-sertation, University of North Dakota, Grand Forks.

Merriam, G. 1995. Movement in spatially divided populations: Responses to landscape structure. In W. Z. Lidicker, Jr., ed., *Landscape approaches in mammalian ecology and conservation.* Minneapolis: University of Min-nesota Press.

Murphy, K. M., G. S. Felzien, and S. E. Relyea. 1992. The ecology of the mountain lion in the northern Yellowstone ecosystem. Cumulative progress report 5. Moscow, Idaho: Wildlife Research Institute.

Myers, R. L., and J. J. Ewel. 1990. *Ecosystems of Florida.* Orlando: University of Central Florida Press.

Nowak, R. M. 1991. *Walker's mammals of the world.* Vols. 1 and 2. Baltimore: Johns Hopkins University Press.

Nowak, R. M., and R. T. McBride. 1973. Feasibility of a study on the Florida panther. Report to the World Wildlife Fund. Mimeo.

———. 1974. Status of the Florida panther. In *World Wildlife Yearbook, 1973–1974.* Morges, Switz.: World Wildlife Fund.

O'Brien, S. J., D. E. Wildt, M. Bush, T. M. Caro, C. Fitzgibbon, I. Aggundy, and R. E. Leakey. 1987. East African cheetahs: Evidence for two population bottlenecks? *Proc. Nat. Acad. Sci. U.S.A.* 84:508–511.

O'Brien, S. J., M. E. Roelke, N. Yuhki, K. W. Richards, W. E. Johnson, W. L. Franklin, A. E. Anderson, O. L. Bass, R. C. Belden, and J. S. Martenson. 1990. Genetic introgression within the Florida panther (*Felis concolor coryi*). *Nat. Geogr. Res. and Explor.* 6:485–494.

Olmstead, I., and L. L. Loope. 1984. Plant communities of Everglades National Park. In P. J. Gleason, ed., *Environments of south Florida: Present and past.* Vol. 2. Coral Gables: Miami Geological Society.

Parker, G. 1995. *Eastern coyote: The story of its success.* Halifax: Nimbus.

Petersen, D. 1995. *Ghost grizzlies.* New York: Holt.

Pritchard, P.C.H. (ed.). 1976. *Proceedings of the Florida Panther Conference.* Orlando: Florida Audubon Society and Florida Game and Fresh Water Fish Commission.

Pulliam, J. 1991. Genetic management, the hybrid policy under the Endangered Species Act, and species concepts. U.S. Fish and Wildlife Service. Unpublished mimeo. Atlanta, Georgia.

Quammen, D. 1996. *The song of the dodo.* New York: Scribner. 702 pp.

Rabb, G. B. 1959. Reproductive and vocal behavior in captive pumas. *J. Mammal.* 40:616–617.

Rabinowitz, A. R., and B. G. Nottingham. 1986. Ecology and behavior of the jaguar (*Panthera onca*) in Belize, Central America. *J. Zool.* (London) 210:149–159.

Regan, T. W., and D. S. Maehr. 1990. Melanistic bobcats in Florida. *Florida Field Nat.* 18:84–87.

Robertson, W. B., Jr., and J. A. Kushlan. 1974. The southern Florida avifauna. In P. J. Gleason II, ed., *Environments of south Florida: Present and past.* Memoir 2. Coral Gables: Miami Geological Society.

Robinette, W. L. 1961. Notes in cougar productivity and life history. *J. Mammal.* 42:204–217.

Robinette, W. L., J. S. Gashwiler, and O. W. Morris. 1959. Food habits of the cougar in Utah and Nevada. *J. Wildl. Manage.* 23:261–273.

Roelke, M. E., J. S. Martenson, and S. J. O'Brien. 1993. The consequences of demographic reduction and genetic depletion in the endangered Florida panther. *Current Biol.* 3:340–350.

Roelke, M. E., E. R. Jacobson, G. V. Kollias, and D. J. Forrester. 1985. Medical management and biomedical findings on the Florida panther, *Felis concolor coryi,* July 1, 1985 to June 30, 1986. Annual report. Tallahassee: Florida Game and Fresh Water Fish Commission.

Roelke, M. E., D. J. Forrester, E. R. Jacobson, G. V. Kollias, F. W. Scott, M. C. Barr, J. F. Evermann, and E. C. Pirtle. 1993. Seroprevalence of infectious disease agents in free-ranging Florida panthers (*Felis concolor coryi*). *J. Wildl. Dis.* 29:36–49.

Roof, J. C., and D. S. Maehr. 1988. Sign surveys for panthers on peripheral areas of their known range. *Florida Field Nat.* 16:81–85.

Schaller, G. B. 1972. *The Serengeti lion: A study of predator–prey relations.* Chicago: University of Chicago Press.

Schaller, G. B., and P. G. Crawshaw, Jr. 1980. Movement patterns of jaguar. *Biotropica* 12:161–168.

Schaller, G. B., T. Qitao, K. G. Johnson, W. Xiaoming, S. Heming, and H. Jinchu. 1989. The feeding ecology of giant pandas and Asiatic black bears in the Tangjiahe Reserve, China. In J. L. Gittleman, ed., *Carnivore behavior, ecology, and evolution.* Ithaca: Cornell University Press.

Schemnitz, S. D. 1974. Populations of bear, panther, alligator and deer in the Florida Everglades. *Florida Scientist* 37:156–167.

Schortemeyer, J. L., D. S. Maehr, J. W. McCown, E. D. Land, and P. D. Manor. 1991. Prey management for the Florida panther: A unique role for managers. *Trans. N. Am. Wildl. Conf.* 56:512–526.

Seal, U. S. (ed.). 1992. Genetic management strategies and population viability of the Florida panther (*Felis concolor coryi*). Report of a workshop to the U.S. Fish and Wildlife Service. Apple Valley, Minn.: Captive Breeding Specialist Group, IUCN.

Seal, U. S., and R. Lacy. 1989. Florida panther population viability analysis. Report to the U.S. Fish and Wildlife Service. Apple Valley, Minn.: Captive Breeding Specialist Group, IUCN.

Seidensticker, J. C., IV, M. G. Hornocker, W. V. Wiles, and J. P. Messick. 1973. Mountain lion social organization in the Idaho Primitive Area. *Wildl. Monogr.* 35:1–60.

Shaw, H. 1979. *Mountain lion field guide.* Spec. Rep. 9. Phoenix: Arizona Game and Fish Dept.

———. 1989. *Soul among lions: The cougar as peaceful adversary.* Boulder: Johnson Books.

Smallwood, K. S. 1993. Mountain lion vocalizations and hunting behavior. *Southwestern Naturalist* 38:65–67.

Smith, T. R., and O. L. Bass. 1994. Landscape, white-tailed deer, and the distribution of Florida panthers in the Everglades. In S. Davis and J. Ogden, eds., *Everglades: The ecosystem and its restoration.* Delray Beach, Fla.: St. Lucie Press.

Sunquist, M. E. 1981. The social organization of tigers (*Panthera tigris*) in Royal Chitwan National Park, Nepal. *Smith. Contrib. Zool.* 336:1–98.

Sweanor, L. L. 1990. Mountain lion social organization in a desert environment. M.S. thesis, University of Idaho, Moscow.

Templeton, A. R. 1986. Coadaptation and outbreeding depression. In M. E. Soulé, ed., *Conservation biology: The science of scarcity and diversity.* Sunderland, Mass.: Sinauer Associates, Inc.

Terborgh, J. 1988. The big things that run the world. *Conserv. Biol.* 2:402–403.

Tinsley, J. B. 1970. *The Florida panther.* St. Petersburg: Great Outdoors.

————. 1987. *The puma: Legendary lion of the Americas.* El Paso: Texas Western Press.

Toweill, D. E. 1977. Food habits of cougars in Oregon. *J. Wildl. Manage.* 41:576–578.

Toweill, D. E., C. Maser, L. D. Bryant, and M. L. Johnson. 1988. Reproductive characteristics of eastern Oregon cougars. *Northwest Sci.* 62:147–150.

U.S. Fish and Wildlife Service. 1987. Florida panther recovery plan. Prepared by the Florida Panther Interagency Committee for the U.S. Fish and Wildlife Service, Atlanta.

U.S. Geological Survey. 1993. South Florida satellite map. Denver: U.S. Geological Survey.

Van Dyke, F. G., and R. H. Brocke. 1987. Sighting and track reports as indices of mountain lion presence. *Wildl. Soc. Bull.* 15:251–256.

Van Dyke, F. G., R. H. Brocke, H. G. Shaw, B. B. Ackerman, T. P. Hemker, and F. G. Lindzey. 1986. Reactions of mountain lions to logging and human activity. *J. Wildl. Manage.* 50:95–102.

Wassmer, D. A., D. D. Guenther, and J. N. Layne. 1988. Ecology of the bobcat in south-central Florida. *Bull. Fla. State Museum Biol. Sci.* 33:159–228.

Webb, S. D. 1974. *Pleistocene mammals of Florida.* Gainesville: University Presses of Florida.

————. 1984. Ten million years of mammal extinctions in North America. In P. S. Martin and R. G. Klein, eds., *Quaternary extinctions—a prehistoric revolution.* Tucson: University of Arizona Press.

White, P. A., and D. K. Boyd. 1989. A cougar, *Felis concolor,* kitten killed and eaten by gray wolves, *Canis lupus,* in Glacier National Park, Montana. *Can. Field Nat.* 103:408–409.

Williams, L. E. 1978. Florida panther. In J. N. Layne, ed., *Rare and endangered biota of Florida.* Vol. 1: *Mammals.* Gainesville: University Presses of Florida.

Wilson, P. 1984. Puma predation on guanacos in Torres del Paine National Park, Chile. *Mammalia* 48:515–522.

Wing, E. S. 1965. Early history. In R. F. Harlow and F. K. Jones, eds., *The white-tailed deer in Florida.* Tech. Bull. 9. Tallahassee: Florida Game and Fresh Water Fish Commission.

Yanez, J. L., J. C. Cardenas, P. Gezelle, and F. M. Jaksic. 1986. Food habits of the southernmost mountain lions (*Felis concolor*) in South America: Natural versus livestocked ranges. *J. Mammal.* 67:604–606.

Young, S. P. 1946. History, life habits, economic status, and control. In S. P. Young and E. A. Goldman, eds., *The puma: Mysterious American cat.* Washington, D.C.: American Wildlife Institute.

Young, S. P., and E. A. Goldman. 1946. *The puma: Mysterious American cat.* Washington, D.C.: American Wildlife Institute.

Young, S. P., and H.H.T. Jackson. 1951. *The clever coyote.* Lincoln: University of Nebraska Press.

INDEX

ABOUT THE AUTHOR

David S. Maehr is currently Assistant Professor of Conservation Biology in the Department of Forestry at the University of Kentucky. He obtained a B.S. in Wildlife Management from Ohio State University and an M.S. and Ph.D. in Wildlife Ecology from the University of Florida. From 1985 to 1994 he directed field studies on the Florida panther and other large mammals for the Florida Game and Fresh Water Fish Commission. He has authored over 60 technical papers on a wide range of wildlife subjects and is coauthor of the book, *Florida's Birds*. His research interests focus on large carnivore ecology and the effects of landscape change on the distribution and diversity of wildlife. Maehr lives with his wife, Diane, and two children, Clifton and Erin.